The Geometrical Tolerancing Desk Reference

Creating and Interpreting ISO Standard Technical Drawings

The Geometrical Tolerancing Desk Reference
Creating and Interpreting ISO Standard Technical Drawings

Paul Green
IEng MIED IngEurEta

European Commission Official
Engineering Designer

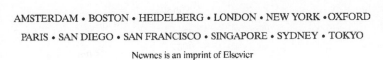

AMSTERDAM • BOSTON • HEIDELBERG • LONDON • NEW YORK • OXFORD
PARIS • SAN DIEGO • SAN FRANCISCO • SINGAPORE • SYDNEY • TOKYO

Newnes is an imprint of Elsevier

ELSEVIER

Newnes

Newnes is an imprint of Elsevier
The Boulevard, Langford Lane, Kidlington, Oxford, OX5 1GB, UK
30 Corporate Drive, Suite 400, Burlington, MA 01803, USA

First edition 2005
Reprinted 2008, 2009

British Library Cataloguing in Publication Data
A catalogue record for this book is available from the British Library

Library of Congress Cataloging-in-Publication Data
A catalog record for this book is available from the Library of Congress

ISBN: 978-0-7506-6821-7

For information on all Newnes publications
visit our website at www.elsevierdirect.com

Printed and bound in the United Kingdom
Transferred to Digital Print 2011

Contents 1

Contents 2

Preface

Geometrical Tolerancing can best be described as a language of symbols placed on technical drawings to adequately define the allowable variation of part geometry.

Geometrical Tolerancing in its various forms has been around for nearly fifty years and is now widely used and accepted throughout manufacturing industry. It has developed to become a valuable tool essential for meeting the high standards demanded by today's modern high technology workplaces. The scope of Geometrical Tolerancing spreads across the board and is a system that can be applied equally to the control of small geometry like electronic components right through to very large geometry used, for example, by companies manufacturing aircraft components.

An accurate and respected communication system of this kind is essential to ensuring that in today's highly competitive global markets a successful end product can be assured giving greater confidence throughout the manufacturing industry.

The current internationally accepted standard in Geometrical Tolerancing is ISO1101: 2004 and is the standard upon which this book is based.

Throughout my many years employed as a Mechanical Engineering Designer I worked in various design and drawing offices. During this time I was unable to find a Geometrical Tolerancing reference book that I could use as a daily desk reference for my work needs thus providing me with the motivation to create this book. I designed the desk reference to fulfil the requirements of anybody who applies or needs to understand Geometrical Tolerancing on technical drawings, this would include Engineering Designers, Draughtsmen, Workshop and Inspection Staff as well as Technicians, Engineers and Students.

PART 1

Introduction and how to use this book

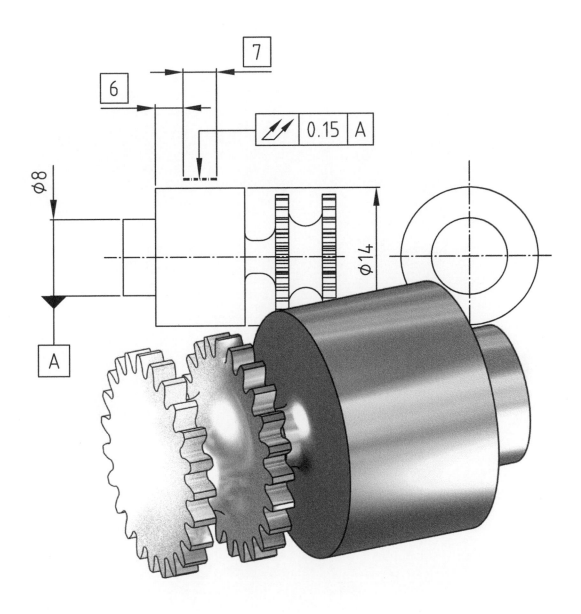

Engineering drawing practice (technical drawing) can be categorized into three sections:

1. General principles

This is concerned with technical drawing layout, scale, line types, text, sections, projection and how to show common features.

2. Dimensioning and size tolerancing

This is concerned with the principles of dimensioning and how to apply size tolerances (not geometrical tolerances) to technical drawings. It also covers machining requirements and surface texture.

3. Geometrical tolerancing

This is concerned with controlling FORM, LOCATION and ORIENTATION of a feature (axis, plane, surface, hole etc...). This is done by applying the principles of geometrical tolerancing to engineering drawings with the aid of geometrical symbols.

This book is concerned with **Geometrical tolerancing** only.

What is a Geometrical Tolerance?

A geometrical tolerance is the maximum allowable variation of form or position of a feature. This is controlled by defining the size and shape of a tolerance zone. The specified part of the feature must be within this tolerance zone.

When to use a Geometrical Tolerance

- The size tolerance of a dimension (see figure 1.3) has a certain amount of control over form and attitude (see figure 1.4), but if a better degree of control is required then geometrical tolerances should be used (see figure 1.5).

- Position of a feature is also controlled by geometrical tolerances.

- The use of geometrical tolerances can increase manufacturing costs, so they should only be used when necessary.

Orthographic Representation

Technical drawings usually consist of various two dimensional views to define an object, this is known as orthogonal projection. The two orthogonal projection methods used internationally are first angle projection and third angle projection. Third angle projection (figure 1.2) is used mainly in The United States and Canada whilst first angle projection (figure 1.1) is used mainly throughout Europe and the rest of the world. Both first and third angle projection have equal status and are approved internationally. In ISO Standard ISO1101: 2004 all figures have been drawn in first angle projection with dimensions and tolerances in millimetres. As this book is based on this standard first angle projection has been used throughout the book.

The identifying symbol for first angle projection ⟨⊏⊐—⊕⟩ or for third angle projection ⟨⊕—⊏⊐⟩ should be shown in the title block of your drawing.

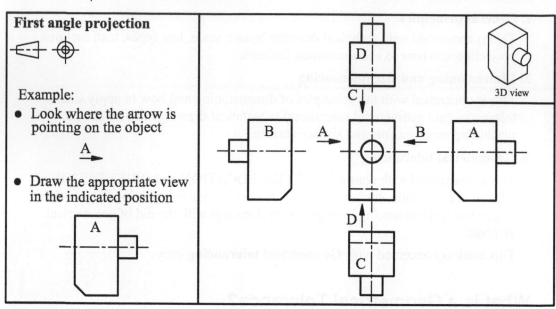

Figure 1.1

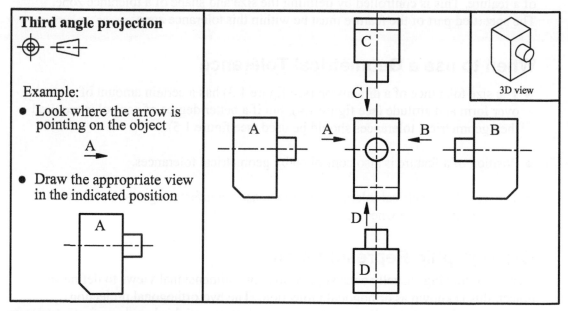

Figure 1.2

Important

- In order to use this book it is essential to read section 2.
- First angle projection is used throughout the book ⟨⊏⊐—⊕⟩

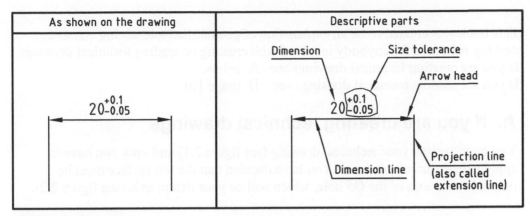

Figure 1.3 Dimension

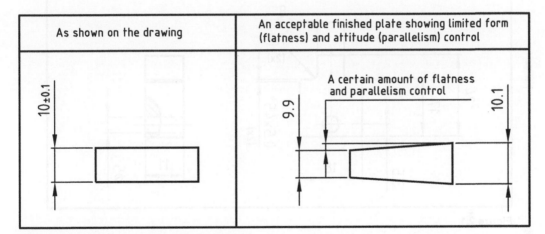

Figure 1.4 Limited form and attitude control

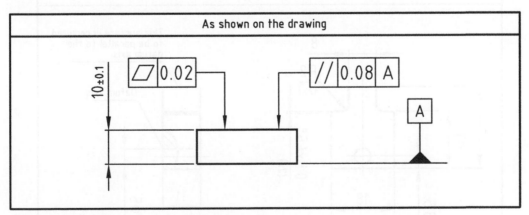

Figure 1.5 Example of geometrical tolerancing

This book was created to be an explanation of geometrical tolerancing and as a desktop reference for anybody involved with creating or reading technical drawings. If you are creating technical drawings see A below. If you are reading technical drawings see B (page 10).

A. If you are creating technical drawings

You have created your technical drawing (see figure 2.1) and now you have to apply a geometrical tolerance. You have decided that the top surface must be parallel to the axis of the Ø5 hole, which will be your datum axis (see figure 2.2).

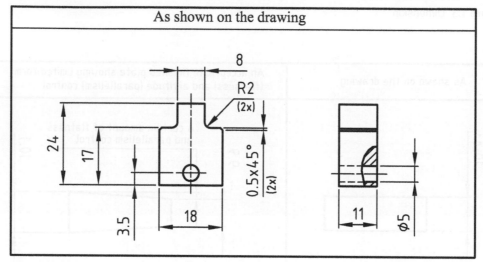

Figure 2.1

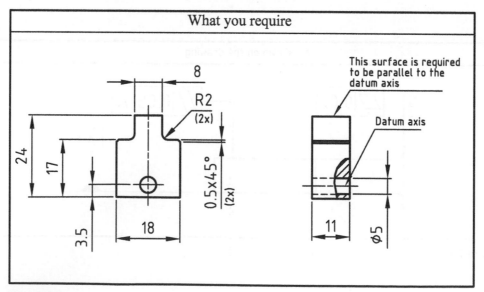

Figure 2.2

Step 1

Refer to section 21 (the visual index, pages 173 to 193) to select a **Parallelism** geometrical tolerance.

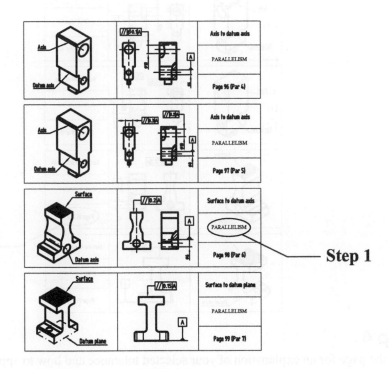

Step 1

Step 2

Second step, for this example you would select a **Surface to datum axis** from the parallelism choices.

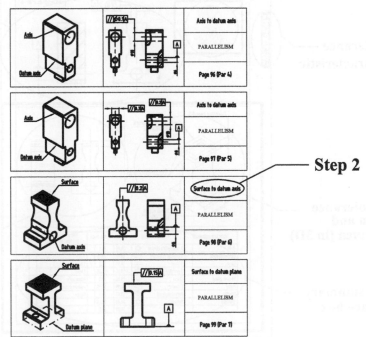

Step 2

Step 3

Read the reference page number to go to next.

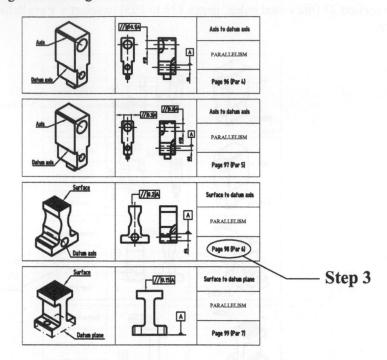

Step 3

Step 4

Read the page for an explanation of your selected tolerance and how to apply it to your drawing.

3D picture

Tolerance characteristic

** How to apply the tolerance to your drawing **

Do not show this information on your drawing

The tolerance shown and explained (in 3D)

The tolerance shown in front and side views (in 2D)

Quick summary reference box

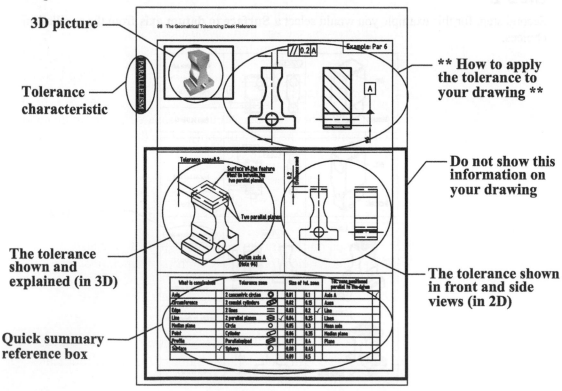

Step 5

Select the amount of tolerance required for your tolerance frame, either from your company's standards or as a guide use table 18.3 on page 63.

Table 18.3 An example of geometrical tolerance values
(normal precision mechanical engineering firm)

Characteristic	Symbol	up to 10	from 10 to 30	from 30 to 100	from 100 to 300	from 300 to 1000	
Angularity	∠	0.02	0.06	0.25	0.8	2	
Circular run-out	⟋	0.05	0.05	0.05	0.05	0.05	
Coaxiality	◎	0.05	0.05	0.1	0.15	0.2	
Concentricity	◎	0.05	0.05	0.1	0.15	0.2	
Cylindricity	⌭	0.25	0.25	0.25	0.3	0.4	
Flatness	▱	0.01	0.02	0.05	0.15	0.3	
Parallelism	//	0.01	0.03	0.1	0.3	1	
Perpendicularity (Squareness)	⊥	0.01	0.03	0.1	0.2	0.5	
Position	⊕	0.05	0.05	0.1	0.15	0.2	
Profile of a line	⌒	0.05	0.05	0.1	0.15	0.2	
Profile of a surface	⌓	0.05	0.05	0.1	0.15	0.2	
Roundness (Circularity)	○	0.1	0.1	0.1	0.1	0.1	
Straightness	—	0.005	0.015	0.05	0.15	0.3	
Total run-out	⟋⟋	0.05	0.05	0.05	0.05	0.05	
Characteristic	**Symbol**	up to 3	from 3 to 6	from 6 to 30	from 30 to 120	from 120 to 400	from 400 to 1000
Symmetry	⌯	0.05	0.1	0.2	0.3	0.5	0.8

Step 6

Apply the geometrical tolerance and datum reference to your drawing.

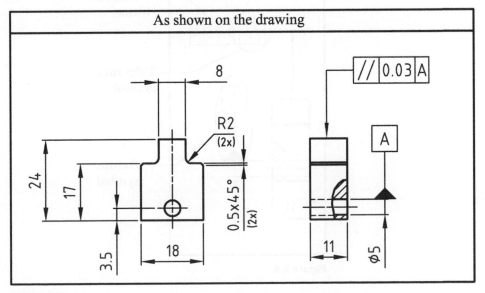

As shown on the drawing

8

R2 (2x)

// 0.03 A

A

24

17

3.5

18

0.5×45° (2x)

11

ø5

Figure 2.3

B. If you are reading technical drawings

You have seen some geometrical tolerancing on a drawing (see figure 2.4) and you don't understand what it means.

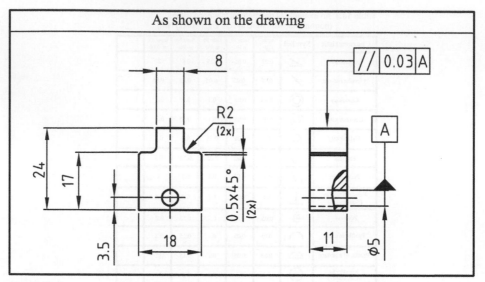

Figure 2.4

Step 1

Look at the geometrical tolerance symbol.

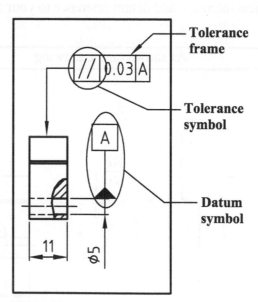

Figure 2.5

Step 2

Look for the geometrical tolerance characteristic given next to the tolerance symbol in table 4.1 (see page 18), in this example it is parallelism.

Table 4.1 Geometric characteristic symbols

Features	Tolerance type	Characteristic	Symbol	Page
For single features (not related to a datum)	Form	Cylindricity		89
		Flatness		90 to 92
		Profile of a line		120 to 124
		Profile of a surface		125 to 128
		Roundness (Circularity)	○	129 to 133
		Straightness	—	134 to 141
For related features (related to a datum)	Orientation (also called Attitude)	Angularity		68 to 74
		Parallelism	//	93 to 102
		Perpendicularity (Squareness)	⊥	103 to 111
		Profile of a line		120 to 124
		Profile of a surface		125 to 128
	Location	Concentricity	◎	87 to 88
		Coaxiality	◎	84 to 86
		Profile of a line		120 to 124
		Profile of a surface		125 to 128
		*Position	⊕	112 to 119
		Symmetry	≐	142 to 147
	Run-out (also called Composite)	Circular run-out	↗	75 to 83
		Total run-out		148 to 151

Tolerance characteristic

Tolerance symbol

* May also not be related to a datum.

Step 3

Refer to section 21 (the visual index, pages 173 to 193) to select a **Parallelism** geometrical tolerance with the same characteristics as shown on the drawing that you are reading i.e. Surface to datum axis.

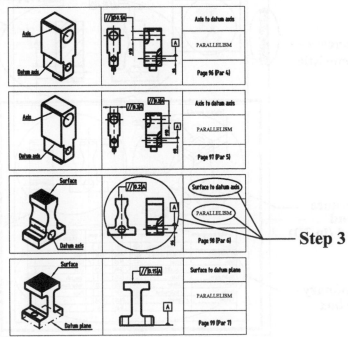

Step 3

Step 4

Read the reference page number to go to next.

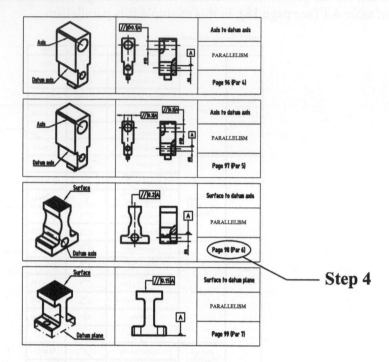

— Step 4

Step 5

Read the page for an explanation of your selected tolerance.

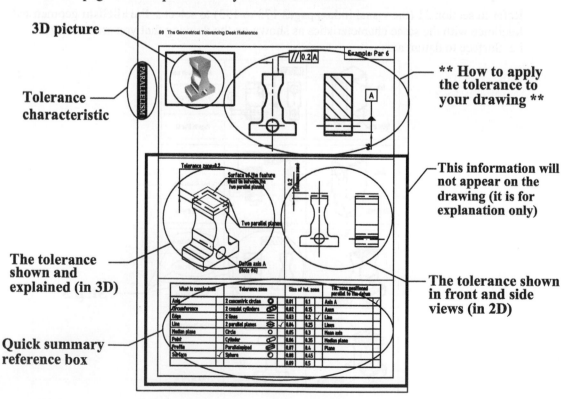

3D picture

Tolerance characteristic

The tolerance shown and explained (in 3D)

Quick summary reference box

** How to apply the tolerance to your drawing **

This information will not appear on the drawing (it is for explanation only)

The tolerance shown in front and side views (in 2D)

Step 6

For an explanation about the **datum symbol** refer to pages 28 to 32.

Basic explanation

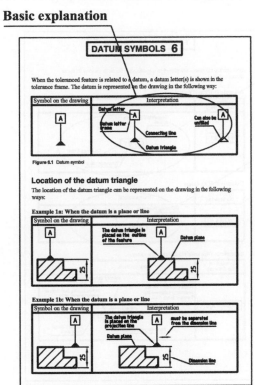

The datum is an axis

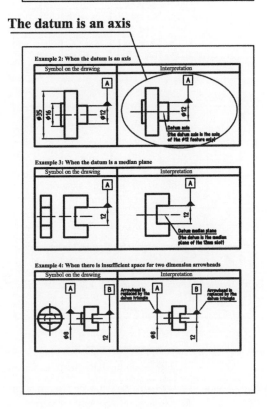

Step 7

For an explanation about the **tolerance frame** refer to pages 23 to 27.

Basic explanation

Single datum reference

(In figure 2.4, a cylindrical tolerance zone is not required)

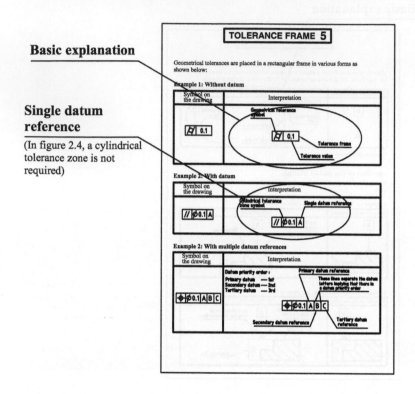

PART 2

Key aspects of geometrical tolerancing

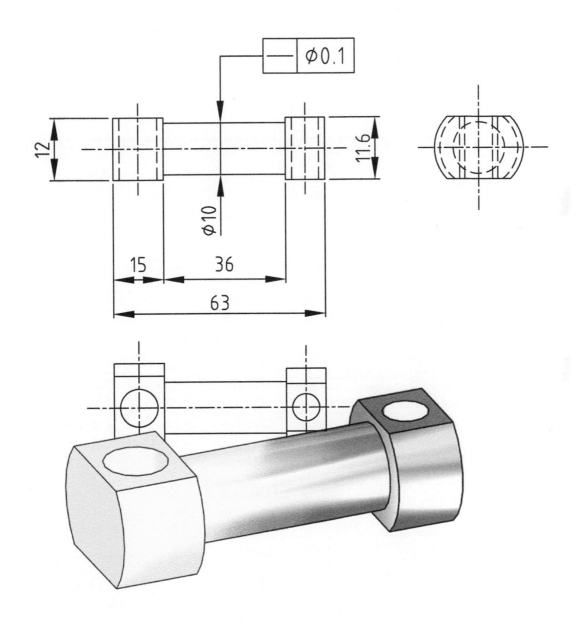

The following sketches illustrate some of the single features that could be on a component.

Some examples of single features
An axis
A cylindrical surface
A cylindrical surface of a hole
An edge
A face
A line on a surface
A median plane
A spherical surface

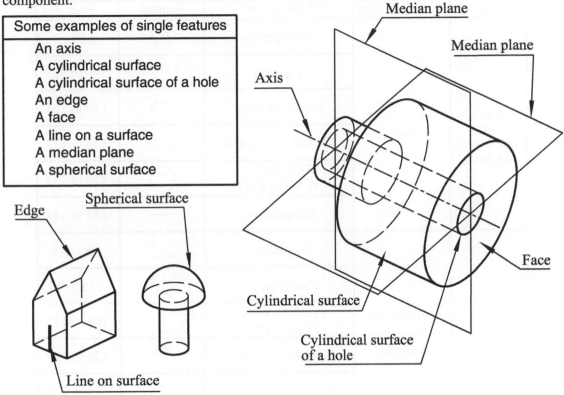

Combinations of Single Features

The following sketch illustrates some combinations of single features that could be on a component.

Some examples of combinations of single features
A groove
A slot
A tongue

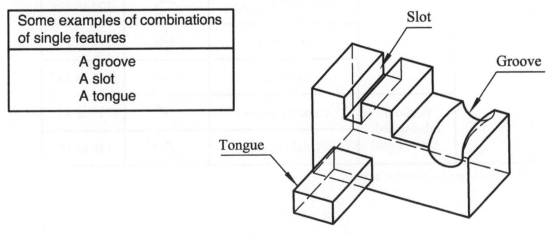

Table 4.1 Geometric characteristic symbols

Features	Tolerance type	Characteristic	Symbol	Page
For single features (not related to a datum)	Form	Cylindricity		89
		Flatness		90 to 92
		Profile of a line		120 to 124
		Profile of a surface		125 to 128
		Roundness (Circularity)		129 to 133
		Straightness		134 to 141
For related features (related to a datum)	Orientation (also called Attitude)	Angularity		68 to 74
		Parallelism		93 to 102
		Perpendicularity (Squareness)		103 to 111
		Profile of a line		120 to 124
		Profile of a surface		125 to 128
	Location	Concentricity		87 to 88
		Coaxiality		84 to 86
		Profile of a line		120 to 124
		Profile of a surface		125 to 128
		*Position		112 to 119
		Symmetry		142 to 147
	Run-out (also called Composite)	Circular run-out		75 to 83
		Total run-out		148 to 151

* May also not be related to a datum.

Table 4.2 Symbols for geometrical tolerancing

Symbol on the drawing	Description	Interpretation
⊕ ⌀0.1 A	Tolerance frame (also known as a feature control frame)	Geometrical tolerance symbol — Datum letter — ⊕ ⌀0.1 A — Tolerance frame — Cylindrical tolerance zone symbol — Tolerance value
6x ⊕ ⌀0.1 A	Multiple tolerance frames	Number of toleranced features — 6x ⊕ ⌀0.1 A
(toleranced feature indicator symbol)	Toleranced feature indicator	Toleranced feature indication direct — Toleranced feature indicator
A (datum indicator symbol)	Datum indicator	Datum letter — Datum frame — A — Datum feature indicator (Datum triangle)
(tolerance for axis symbol) 16	Tolerance for an AXIS or a MEDIAN PLANE	Toleranced feature indication direct — Placed on extended dimension line — 16 — Dimension line

Table 4.3 Symbols for geometrical tolerancing

Symbol on the drawing	Description	Interpretation
	Datum is an AXIS or MEDIAN PLANE	Datum indicator — A — Placed on extended dimension line — Dimension line — 16
	Tolerance for a SURFACE or GENERATOR LINE	Toleranced feature indication direct — Placed on projection line — Projection line (extension line) — 16
	Datum is a SURFACE or GENERATOR LINE	Datum indicator — A — Placed on projection line — Projection line (extension line) — 16
A	Toleranced feature indicator	Toleranced feature indication by letter — A — [* Obsolete * Shown for reference to drawings prepared to earlier standards.] — Toleranced feature indicator

Table 4.4 Symbols for geometrical tolerancing

Symbol on the drawing	Description	Interpretation
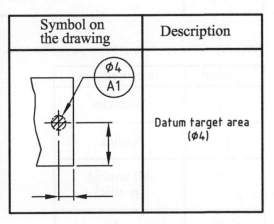	Datum target	Datum target size / Datum target number / Datum identifying letter

Symbol on the drawing	Description	Symbol on the drawing	Description
	Datum target point		Datum target line (front view)
	Datum target line (side view)		Datum target area (Ø4)

Table 4.5 Geometric characteristic symbols

Symbol	Interpretation	Refer to
20	Theoretically exact dimensions (TED) (sometimes called true or boxed dimensions)	Section 10
(P)	Projected tolerance zone	Section 11
(M)	Maximum material requirement	Section 12
(L)	Least material requirement	Section 13
(E)	Envelope requirement	Section 14
(F)	Free state condition (non-rigid parts)	Section 15
CZ	Common zone	
LD	Minor diameter	
MD	Major diameter	
PD	Pitch diameter	
LE	Line element	
NC	Not convex	
ACS	Any cross-section	

Table 4.6 Other symbols used throughout the book

Symbol	Interpretation
ϕ	Diameter
R	Radius
	All around (profile)
Sϕ	Sphere diameter

Geometrical tolerances are placed in a rectangular frame in various forms as shown below:

Example 1: Without datum

Symbol on the drawing	Interpretation
◿ 0.1	Geometrical tolerance symbol → ◿ 0.1 ← Tolerance frame; Tolerance value

Example 2: With datum

Symbol on the drawing	Interpretation
// ⌀0.1 A	Cylindrical tolerance zone symbol; Single datum reference → // ⌀0.1 A

Example 3: With multiple datum references

Symbol on the drawing	Interpretation
⊕ ⌀0.1 A B C	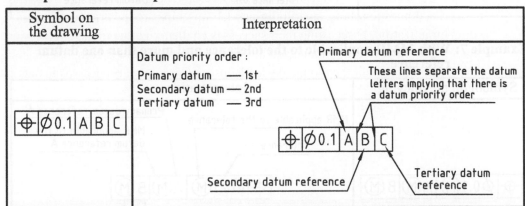

Datum priority order :

Primary datum — 1st
Secondary datum — 2nd
Tertiary datum — 3rd

Primary datum reference

These lines separate the datum letters implying that there is a datum priority order

Secondary datum reference

Tertiary datum reference

Example 4: With MMR (maximum material requirement) applicable to the tolerance

Symbol on the drawing	Interpretation
⊕ |⌀0.1 Ⓜ| A |	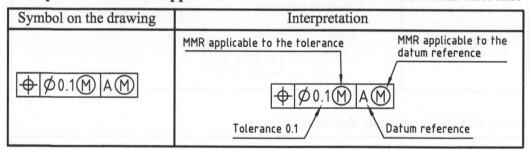

Example 5: With MMR applicable to the datum reference

Symbol on the drawing	Interpretation
⊕ |⌀0.1| A Ⓜ |	

Example 6: With MMR applicable to both the tolerance and the datum reference

Symbol on the drawing	Interpretation
⊕ |⌀0.1Ⓜ| A Ⓜ |	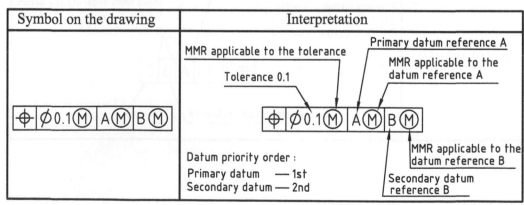

Example 7: With MMR applicable to the tolerance and more than one datum reference

Symbol on the drawing	Interpretation
⊕|⌀0.1Ⓜ|AⓂ|BⓂ|	MMR applicable to the tolerance Tolerance 0.1 Primary datum reference A MMR applicable to the datum reference A ⊕|⌀0.1Ⓜ|AⓂ|BⓂ| Datum priority order : Primary datum —1st Secondary datum — 2nd MMR applicable to the datum reference B Secondary datum reference B

Example 8: When the tolerance is restricted to a length of a feature

Symbol on the drawing	Interpretation
// 0.02/50 A	Note : Restricted length is ANY measured length (in this case of 50mm long) in the total length of the feature Restricted length // 0.02/50 A Tolerance 0.02

Example 9: When the tolerance is restricted to the whole feature and a length of the feature

Symbol on the drawing	Interpretation
// 0.1 / 0.02/50 A	Note : 0.1 tolerance applies to the whole length and 0.02 tolerance applies to any 50mm length within the whole feature length Tolerance 0.1 for the whole feature // 0.1 / 0.02/50 A Tolerance 0.02 Restricted length

Example 10: With MMR applicable to the tolerance and more than one datum reference (the sequence of either datum is of no importance)

Symbol on the drawing	Interpretation
⊕ ⌀0.1Ⓜ AⓂ B Ⓜ	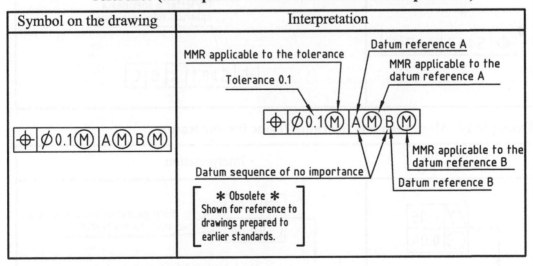

Example 11: When the tolerance applies to more than one feature

Symbol on the drawing	Interpretation
(4x) ⌀ 0.1	The number of features the tolerance applies to (4x) ⌀ 0.1

Example 12: To specify the form of the feature within the tolerance zone

Symbol on the drawing	Interpretation
▱ 0.06 NC	▱ 0.06 NC The form of the feature (in this case NC = Not convex)

Example 13: To specify a spherical tolerance zone

Symbol on the drawing	Interpretation
⊕ S⌀0.1 A B C	The letter S specifies a spherical tolerance zone ⊕ S⌀0.1 A B C

Example 14: More than one tolerance frame for one feature

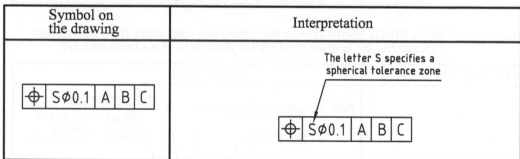

Symbol on the drawing	Interpretation
⌀ 0.06 ◯ 0.04	⌀ 0.06 ◯ 0.04 — Both geometrical characteristics apply to one feature The toleranced feature indicator points to the required feature

Example 15: When the form of the feature is a line instead of a surface

Symbol on the drawing	Interpretation
//|0.06|A|B LE	//|0.06|A|B LE The letters LE mean line element / The 0.06 tolerance applies to a line instead of a surface

Example 16: A single tolerance zone applying to several separate features (common zone)

Symbol on the drawing	Interpretation
⟋|0.04 CZ	⟋|0.04 CZ The letters CZ mean common zone / All features must be within the common tolerance zone of 0.4

Example 17: A profile characteristic applying to the entire outline of the cross-sections

Symbol on the drawing	Interpretation
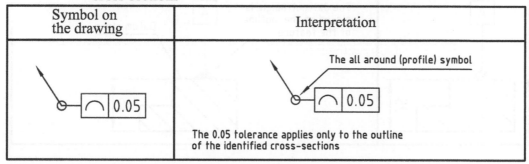	The all around (profile) symbol ⌒|0.05 The 0.05 tolerance applies only to the outline of the identified cross-sections

Example 18: A profile characteristic applying to the entire surface

Symbol on the drawing	Interpretation
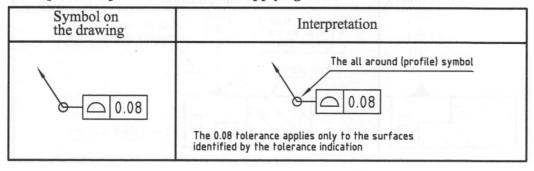	The all around (profile) symbol ⌓|0.08 The 0.08 tolerance applies only to the surfaces identified by the tolerance indication

When the toleranced feature is related to a datum, a datum letter(s) is shown in the tolerance frame. The datum is represented on the drawing in the following way:

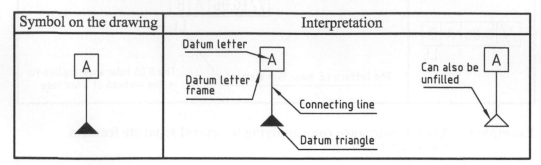

Figure 6.1 Datum symbol

Location of the datum triangle

The location of the datum triangle can be represented on the drawing in the following ways:

Example 1a: When the datum is a plane or line

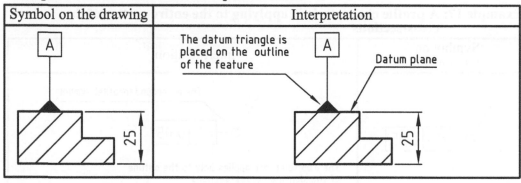

Example 1b: When the datum is a plane or line

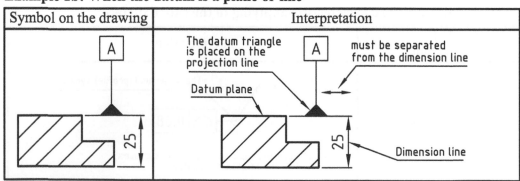

Example 2: When the datum is an axis

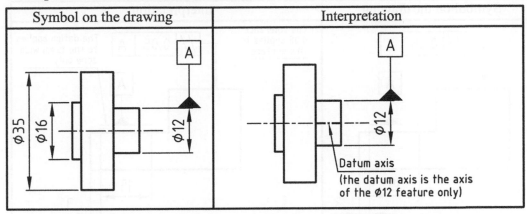

Symbol on the drawing	Interpretation

Example 3: When the datum is a median plane

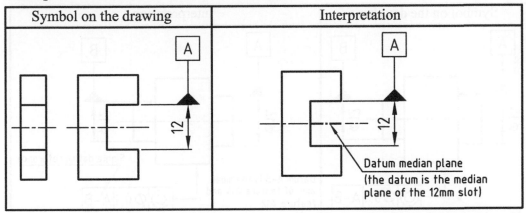

Symbol on the drawing	Interpretation

Example 4: When there is insufficient space for two dimension arrowheads

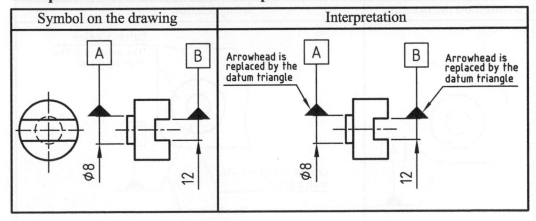

Symbol on the drawing	Interpretation

Example 5: When the datum is applied to a restricted part of a feature

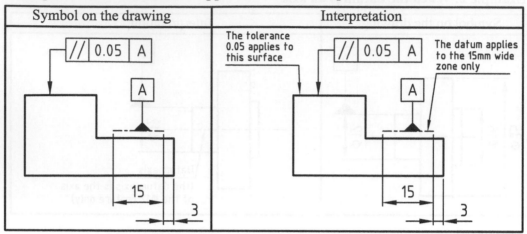

Symbol on the drawing	Interpretation

Example 6: When two datum features combine to make one single datum reference

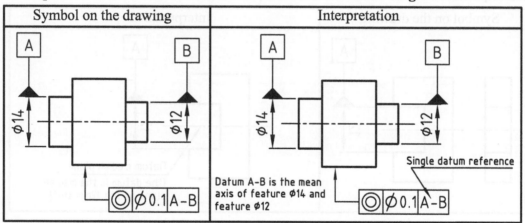

Symbol on the drawing	Interpretation

Example 7: When the datum is a surface in the drawing plane

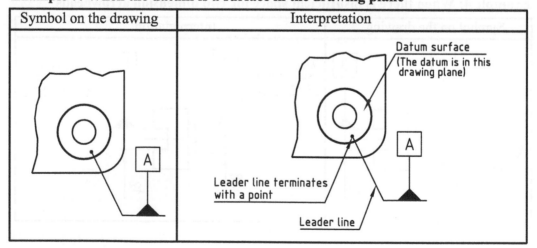

Symbol on the drawing	Interpretation

Example 8a: When the datum is one common axis (or plane) formed by two different features

Symbol on the drawing	Interpretation

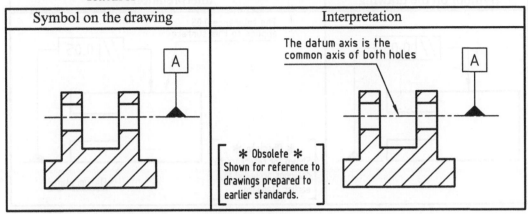

Example 8b: When the datum is one common axis (or plane)

Symbol on the drawing	Interpretation

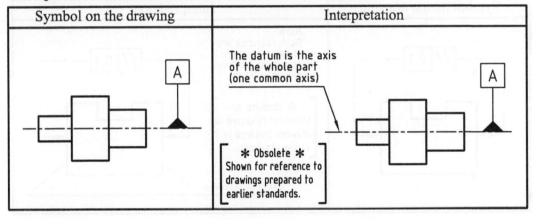

Example 9a: When the datum is without a datum letter

Symbol on the drawing	Interpretation

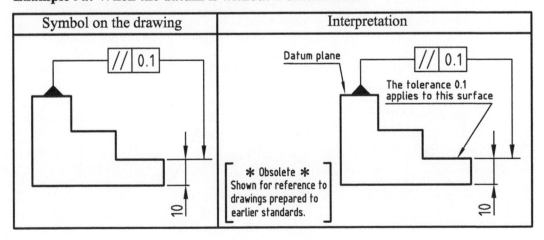

Example 9b: When the datum is without a datum letter

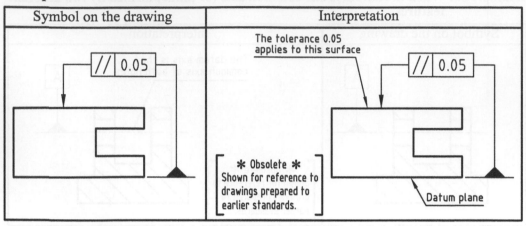

Symbol on the drawing	Interpretation

Example 10: Toleranced feature and datum are the same

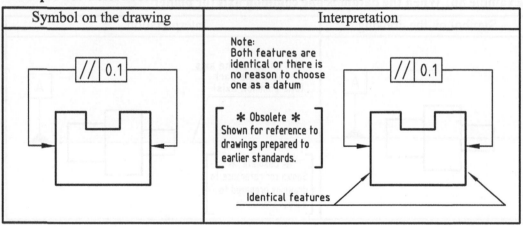

Symbol on the drawing	Interpretation

A datum target is either a point, line or area on a workpiece. The datum target is to be the place of contact (supports of the workpiece) for the manufacturing and inspection equipment.

When a part has an irregular contour (casting, forging, sheet metal parts etc...) the use of a datum target could be required.

Example 1: Interpretation of a datum target

Symbol on the drawing	Description	Interpretation
⌀6 / A1	Datum target	⌀6 / A1 — Datum target size, Datum target number, Datum identifying letter

Example 2a: Type of datum target

Symbol on the drawing	Description	Interpretation	

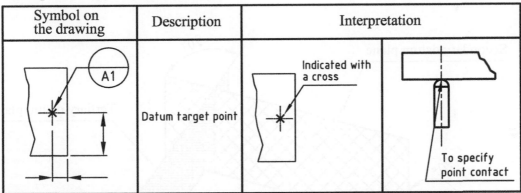

Datum target point — Indicated with a cross — To specify point contact

Example 2b: Type of datum target

Symbol on the drawing	Description	Interpretation	

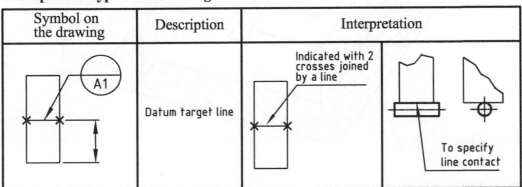

Datum target line — Indicated with 2 crosses joined by a line — To specify line contact

Example 2c: Type of datum target

Symbol on the drawing	Description	Interpretation	
⌀6 A1	Datum target area (⌀6)	Indicated with a hatched area within a double dashed chain line	⌀6 To specify area contact (can be any shape)

Three-plane Datum System

Normally for orientation tolerances the use of one datum or two are sufficient.
Sometimes when using positional relationships a three-plane datum system is required.
In a three-plane datum system all the planes are perpendicular to each other.

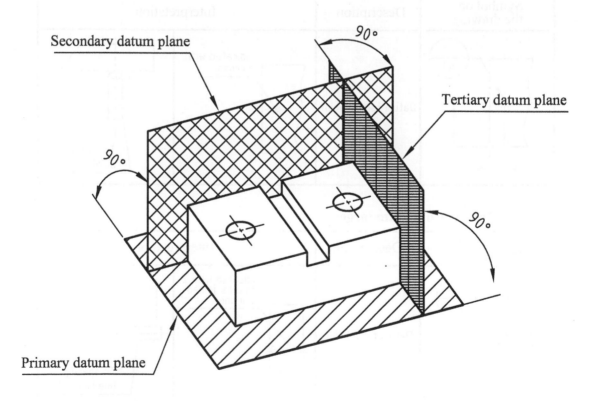

Secondary datum plane

Tertiary datum plane

90°

90°

90°

Primary datum plane

Figure 7.1 The three-plane datum system

Example 3: As shown on the drawing

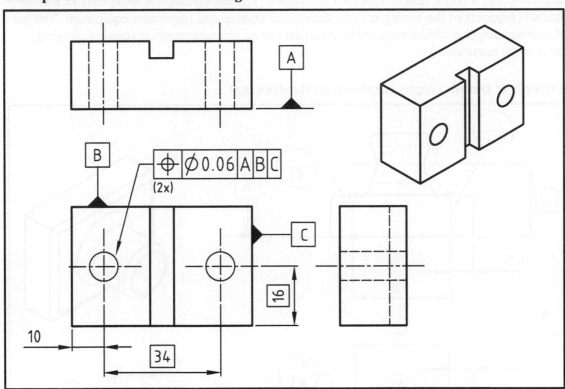

Note: The datums have a priority order.

Datum priority order :

Primary datum — 1st
Secondary datum — 2nd
Tertiary datum — 3rd

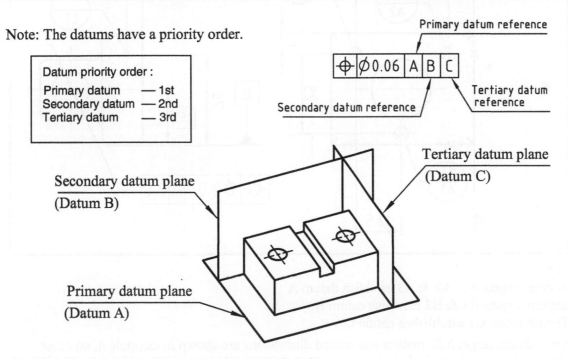

Figure 7.2 The three-plane datum system from example 3 above

Another example of a three-plane datum system is by the use of datum targets. The datum target is either a point, line or area on a workpiece. The datum target is to be the place of contact (supports of the workpiece) for the manufacturing and inspection equipment. The use of a datum target could be required when a part has an irregular contour (casting, forging, sheet metal parts etc...) .

Example 4: Datum targets (as shown on the drawing)

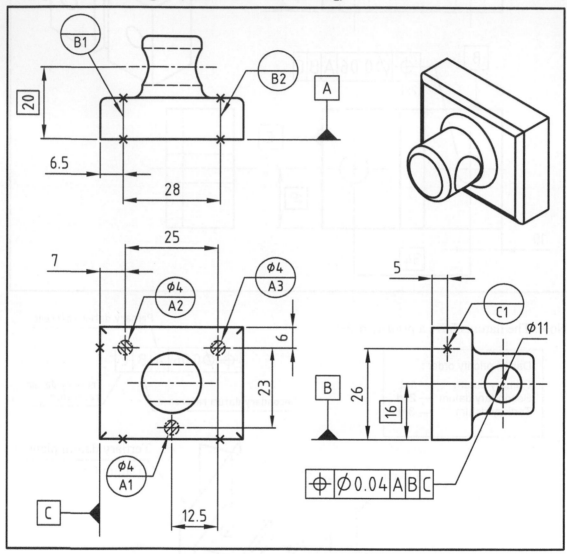

Notes:

Datum targets A1, A2 & A3 establish datum A

Datum targets B1 & B2 establish datum B

Datum target C1 establishes datum C

Only datum target information and related dimensions are shown in example 4, no other dimensions have been included.

TOLERANCED FEATURE **8**

A leader line and arrowhead connects the tolerance frame to the toleranced feature.
The toleranced feature is labelled in the following ways:

Example 1a: Placed on the outline of a feature

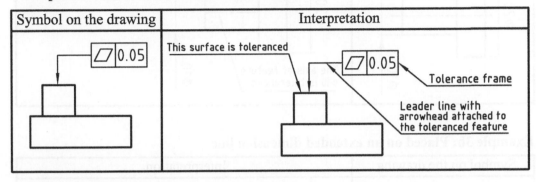

Example 1b: Placed on the outline of a feature

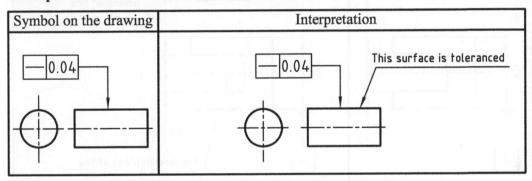

Example 2: Placed on the projection line

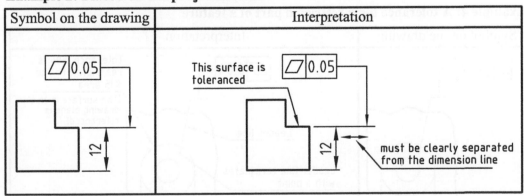

Example 3a: Placed on an extended dimension line

Symbol on the drawing	Interpretation
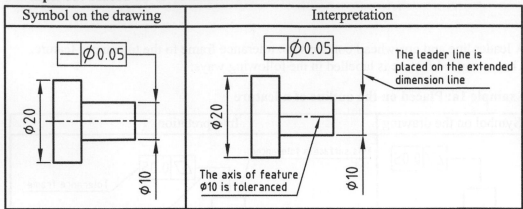	

Example 3b: Placed on an extended dimension line

Symbol on the drawing	Interpretation
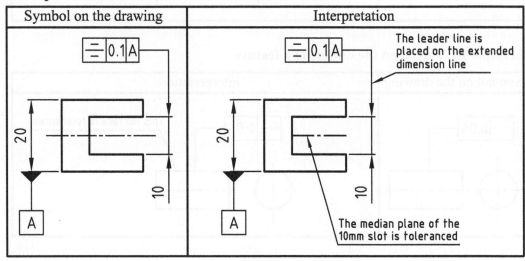	

Example 4: A tolerance for a restricted part of a feature

Symbol on the drawing	Interpretation
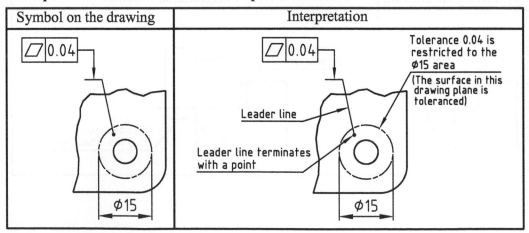	

Example 5: Placed on a reference line

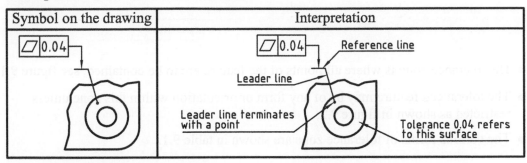

Symbol on the drawing	Interpretation

Example 6: Placed on the axis

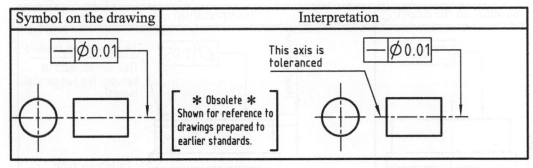

Symbol on the drawing	Interpretation

Example 7a: Placed on a common axis

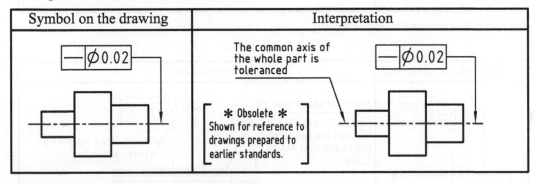

Symbol on the drawing	Interpretation

Example 7b: Placed on a common axis

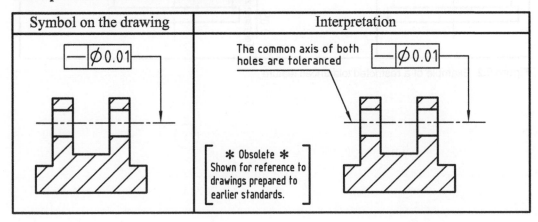

Symbol on the drawing	Interpretation

TOLERANCE ZONE 9

- The tolerance zone is where all points of the feature are to be contained, see figure 9.1.

- The toleranced feature may be of any form or orientation within this zone unless restricted as shown in figure 9.2.

- The various forms of tolerance zone are shown in table 9.1.

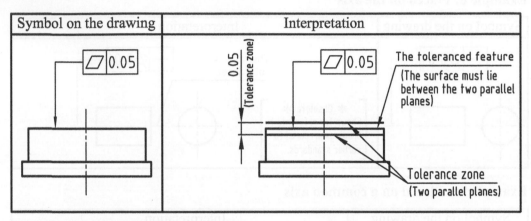

Figure 9.1 Example of a toleranced feature

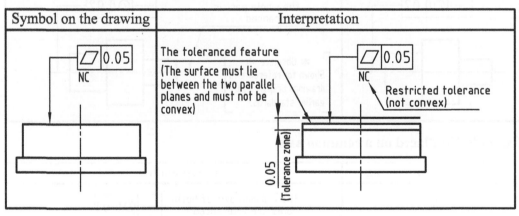

Figure 9.2 Example of a restricted toleranced feature

Table 9.1 Tolerance zones

Tolerance zone	Pictorial representation
Inside a circle	
Between two concentric circles	
Between two equidistant lines OR Between two parallel straight lines	
Inside a sphere	
Inside a cylinder	
Between two coaxial cylinders	
Between two parallel planes OR Between two equidistant surfaces	
* Inside a parallelepiped	

* Note:
Inside a parallelepiped is equivilent to the space between two horizontal planes and two vertical planes.

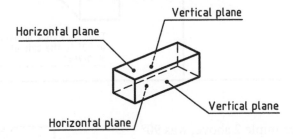

Tolerance Zone Direction

Tolerance zone width is always NORMAL to the surface (see example 1 below) unless otherwise indicated (see example 2 below).

Example 1: Tolerance zone normal to the surface

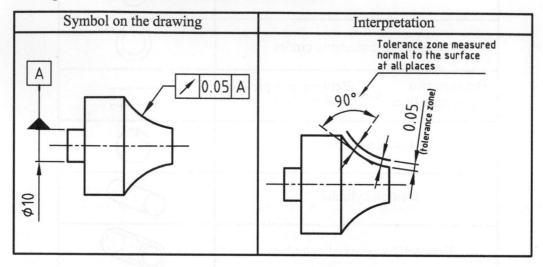

Example 2: Tolerance zone at a specified angle to the surface

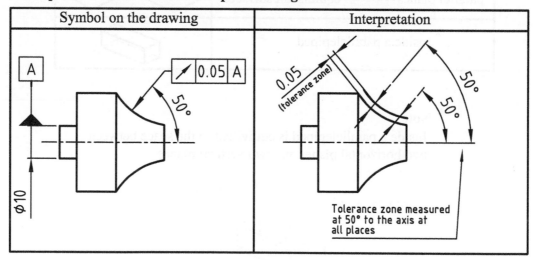

Note:
If the angle shown in example 2 above, was 90° (instead of 50°) as in figure 9.3, it would still be shown on the drawing. This is not the same as normal to the surface as in example 1. In example 1 the 90° angle follows the contour of the surface, in example 2 the angle is always fixed in relation to the axis.

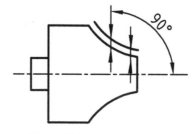

Figure 9.3

Example 3: The same tolerance value for more than one feature

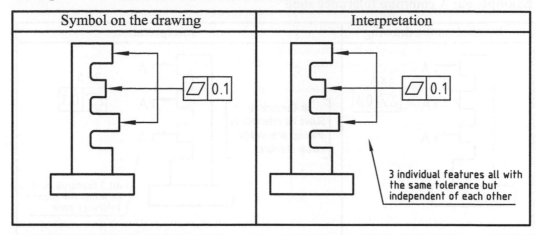

Symbol on the drawing	Interpretation

Example 4: Restricted tolerance

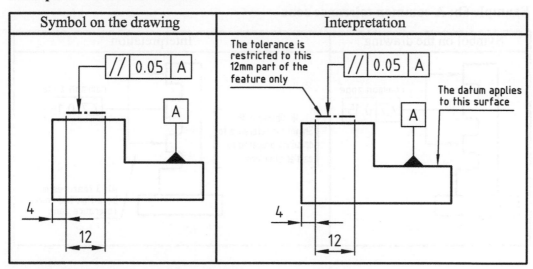

Symbol on the drawing	Interpretation

Example 5: A common tolerance zone

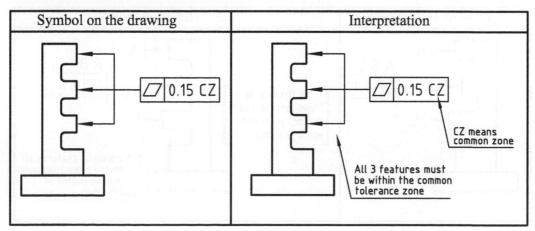

Symbol on the drawing	Interpretation

Example 6a: A common tolerance zone

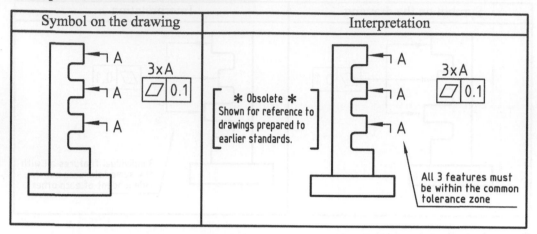

Example 6b: A common tolerance zone

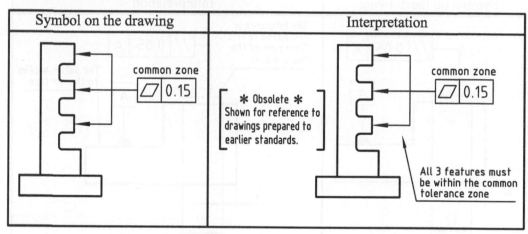

Example 7: The same tolerance value for more than one feature

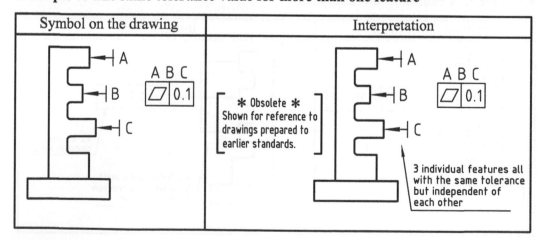

THEORETICALLY EXACT DIMENSIONS **10**

Theoretically exact dimensions (TED), also known as true or boxed dimensions must not be toleranced. The dimension is shown in a rectangular frame.

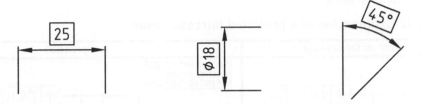

Theoretically exact dimensions may only vary by the geometrical tolerance that is stated in the tolerance frame associated with them.

Example: ⌒ 0.04   by 0.04 only

Theoretically exact dimensions should be used when dimensioning the theoretically exact location of features for tolerances of Angularity, Position, Profile of a line and Profile of a surface.

Example 1: Use of theoretically exact dimensions

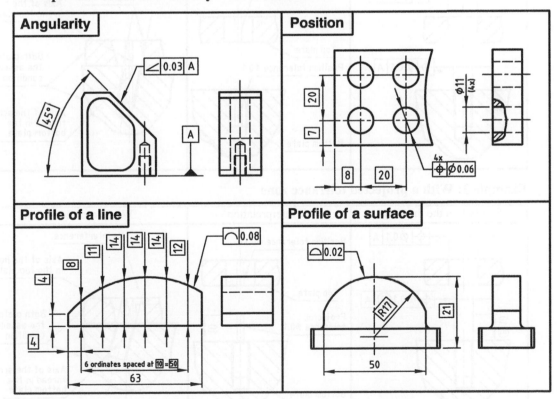

Projected tolerance zone symbol Ⓟ
In the case of a tolerance of orientation (angularity) and position, the tolerance can also apply to the external projection of it and NOT to the tolerance itself, hence this is a projected tolerance zone.

Example 1: Explanation of a projected tolerance zone

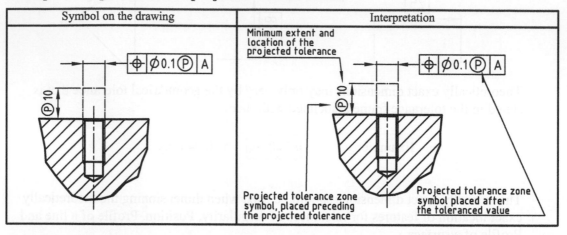

Example 2: No projected tolerance zone

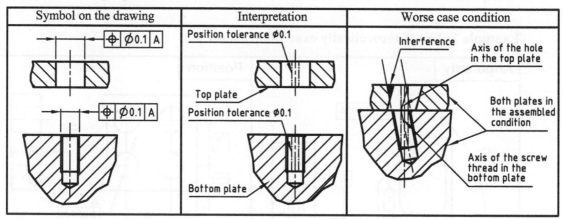

Example 3: With a projected tolerance zone

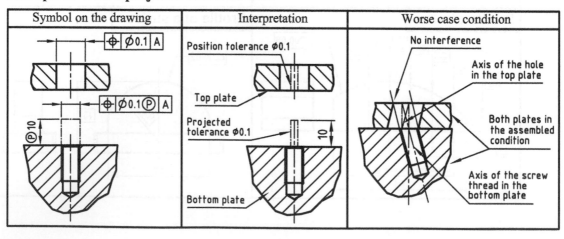

Maximum material condition (MMC) is the condition when a component or feature of a component has the maximum amount of material.

Example 1:
For a shaft, MMC would be when the diameter is everywhere at its maximum size.

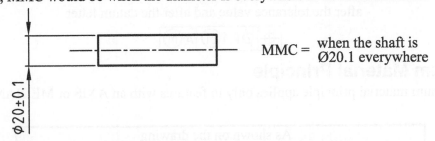

MMC = when the shaft is Ø20.1 everywhere

For a hole, MMC is when its diameter is everywhere at its minimum size.

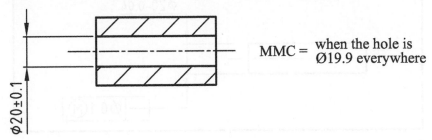

MMC = when the hole is Ø19.9 everywhere

Table 12.1 MMC related definitions

Definition	Abbreviation	Meaning
Maximum material condition	MMC	The state of a feature where the feature is everywhere at its maximum material
Maximum material size	MMS	The limit of size where the material of a feature is at its maximum. Shaft - maximum limit of size Hole - minimum limit of size
Maximum material virtual size	MMVS	The maximum material size plus or minus the geometrical tolerance. (for a shaft) MMVS=MMS + geometrical tolerance (for a hole) MMVS=MMS – geometrical tolerance
Maximum material requirement	MMR	When MMR is required on a drawing it is indicated on the drawing by placing the symbol Ⓜ in the tolerance frame either after the geometrical tolerance, after the datum letter or both

The Ⓜ symbol can be placed in the tolerance frame in the following ways:

- after the tolerance value

| ⊕ | ⌀0.1Ⓜ | A |

- after the datum letter

| ⊕ | ⌀0.1 | AⓂ |

- after the tolerance value and after the datum letter

| ⊕ | ⌀0.1Ⓜ | AⓂ |

Maximum Material Principle

The maximum material principle applies only to features with an AXIS or MEDIAN PLANE.

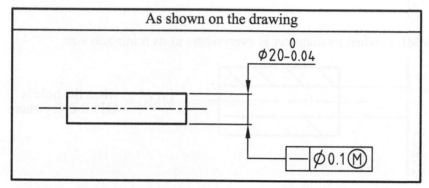

Figure 12.1 Example of applying maximum material principle

The actual measured diameter of the pin (figure 12.1) depicts the total increase in straightness tolerance. If the actual pin is ⌀20 the allowed straightness tolerance is 0.1 but if the actual pin is ⌀19.96 then the allowed straightness tolerance increases to 0.14 (this applies to all other sizes from ⌀19.96 to ⌀20). The straightness tolerance is allowed to increase from 0.1 to a maximum of 0.14 depending on the actual measured size of the pin (see table 12.2).

Table 12.2

Actual measured ⌀ of pin	Additional straightness tolerance allowed	Straightness tol. + Additional straightness tol. = Total geometrical tolerance allowed
20.00 Maximum material condition	0	0.1+ 0 = 0.1
19.99	0.01	0.1+ 0.01 = 0.11
19.98	0.02	0.1+ 0.02 = 0.12
19.97	0.03	0.1+ 0.03 = 0.13
19.96 Least material condition	0.04	0.1+ 0.04 = 0.14

Progressively increasing
tolerance as the pin
⌀ decreases

Note:

Maximum material virtual size = Actual Ø of the measured pin + Maximum straightness tolerance allowed

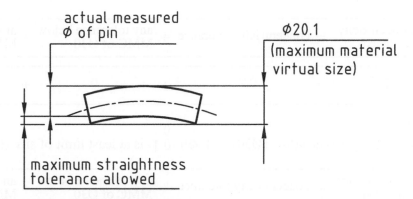

If the pin measures Ø19.97 then the maximum straightness tolerance allowed is 0.13 thus making the maximum material virtual size of Ø20.1

If the pin measures Ø19.99 then the maximum straightness tolerance allowed is 0.11 thus making the maximum material virtual size of Ø20.1

Maximum Material Principle Applied to Datum Features

This principle allows the DATUM AXIS or DATUM MEDIAN PLANE to float relative to the toleranced feature. Further addition to the stated tolerance in the feature frame can be achieved by applying MMR to the datum.

Example 2:

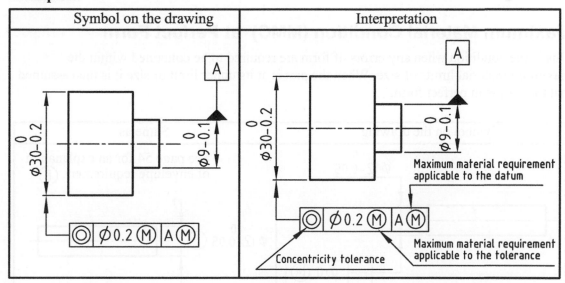

For example 2 the following scenarios could occur:

A. When $\phi 30\!-\!^{0}_{0.2}$ and $\phi 9\!-\!^{0}_{0.1}$ are both at their MMC (Ø30 and Ø9 respectively)

Total concentricity that is allowed	= concentricity tolerance	+	any tolerance below MMC of Ø30	+	any tolerance below MMC of Ø9
0.2	= 0.2	+	0	+	0

B. When $\phi 30\!-\!^{0}_{0.2}$ is at MMC (Ø30) and $\phi 9\!-\!^{0}_{0.1}$ is at least limit of size (Ø8.9)

Total concentricity that is allowed	= concentricity tolerance	+	any tolerance below MMC of Ø30	+	any tolerance below MMC of Ø9
0.3	= 0.2	+	0	+	0.1

C. When $\phi 30\!-\!^{0}_{0.2}$ and $\phi 9\!-\!^{0}_{0.1}$ are both at their least limit of size (Ø29.8 and Ø8.9 respectively)

Total concentricity that is allowed	= concentricity tolerance	+	any tolerance below MMC of Ø30	+	any tolerance below MMC of Ø9
0.5	= 0.2	+	0.2	+	0.1

Maximum Material Condition (MMC) at Perfect Form

This is the condition when any errors of form are required to be contained within the maximum material limits of size. When the part is at its upper limit of size it is then assumed that the part is in perfect form.

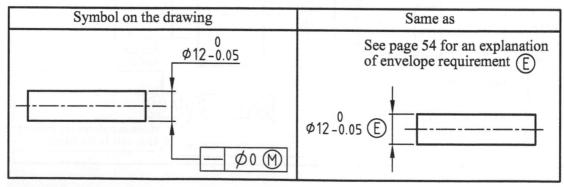

Symbol on the drawing	Same as

Figure 12.2 Example of applying maximum material condition at perfect form

When the pin in figure 12.2 is at its MMC Ø12 the pin must be perfectly straight

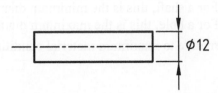

When the pin is at its lowest limit of size Ø11.95 the straightness tolerance can be 0.05

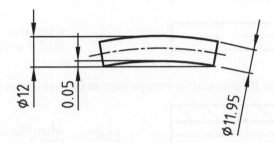

If the pin is at any dimension between Ø12 and Ø11.95 the straightness tolerance is adjusted accordingly i.e. at Ø11.97 the permitted straightness tolerance can be 0.03

Least material condition (LMC) is the condition where the feature of size has the least amount of material within its size limits.

> For a shaft, this is the minimum diameter
> For a hole, this is the maximum diameter

LMC can be used as an alternative to maximum material condition (MMC), it is a condition opposite to MMC.

Examples:

For a shaft, LMC would be when the diameter is everywhere at its minimum size.

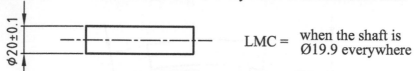

LMC = when the shaft is Ø19.9 everywhere

For a hole, LMC is when its diameter is everywhere at its maximum size.

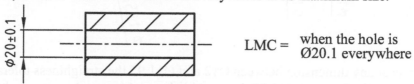

LMC = when the hole is Ø20.1 everywhere

The Ⓛ symbol can be placed in the tolerance frame in the following ways:

after the tolerance value $\boxed{\oplus | \phi 0.1 Ⓛ | A}$, after the datum letter $\boxed{\oplus | \phi 0.1 | A Ⓛ}$, or

after the tolerance value and after the datum letter $\boxed{\oplus | \phi 0.1 Ⓛ | A Ⓛ}$

Table 13.1 LMC related definitions

Definition	Abbreviation	Meaning
Least material condition	LMC	The state of a feature where the feature is everywhere at its minimum material
Least material size	LMS	The limit of size where the material of a feature is at its minimum. Shaft - minimum limit of size Hole - maximum limit of size
Least material virtual size	LMVS	The minimum material size plus or minus the geometrical tolerance. (for a shaft) LMVS=LMS – geometrical tolerance (for a hole) LMVS=LMS + geometrical tolerance
Least material requirement	LMR	When LMR is required on a drawing it is indicated on the drawing by placing the symbol Ⓛ in the tolerance frame either after the geometrical tolerance, after the datum letter or both

Least material principle allows the datum axis (or datum median plane) to float relative to the toleranced feature when leaving least material condition and approaching maximum material condition.

The surface of the datum feature must not violate the least material virtual condition (geometrical ideal form and of least material virtual size).

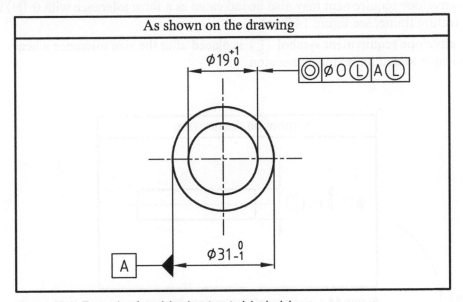

Figure 13.1 Example of applying least material principle

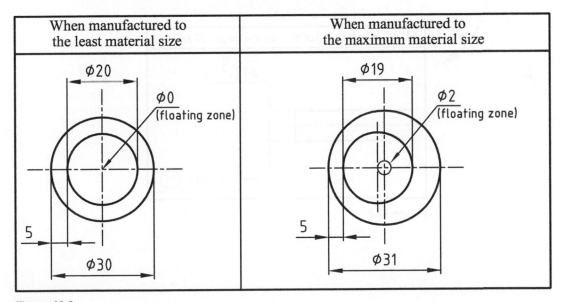

Figure 13.2

Note:

In the example figure 13.1 above, the minimum wall thickness of 5mm for the tube has been achieved when the piece has been manufactured to either the least material size OR the maximum material size.

ENVELOPE REQUIREMENT 14

The envelope requirement represented by symbol Ⓔ specifies that the cylindrical surface of a single feature of size (or feature with two parallel opposite plane surfaces), when at maximum material size, must be within the imaginary envelope of perfect form.

The envelope requirement may also be indicated as a form tolerance with 0 Ⓜ in the feature frame, see figure 14.2 below.

The envelope requirement symbol Ⓔ is placed after the size tolerance when applicable to that individual dimension.

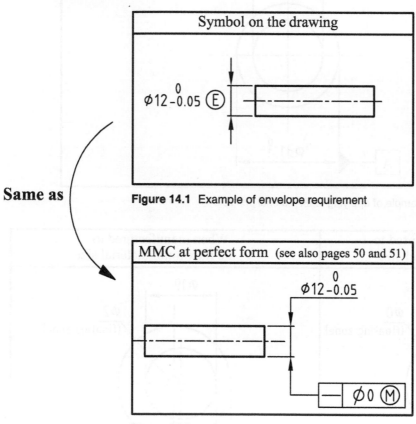

Figure 14.1 Example of envelope requirement

Same as

Figure 14.2

When a part is in a free state condition it is in the unassembled condition, when no external forces are mishaping the part.

Some examples of non-rigid parts are:

        Rubber parts, such as o-rings.
        Plastic parts, such as thin plastic sheets.
        Thin walled parts, such as thin cylinders.

A non-rigid part (flexible part) will have certain dimensions when assembled but will adopt other dimensions when in a free state.

To ensure that a non-rigid part will have the correct dimensions when assembled and functioning, some of the tolerances of the part will apply while the part is restrained, this restraint will represent the restraining forces on the part when it is assembled and functioning.

Some geometric tolerances will apply only when in a free state. This is done by placing the Ⓕ symbol after the tolerance in the tolerance frame i.e. | ◯ | 1 | Ⓕ |.

When "ISO 10579-NR" is placed on the drawing, all geometrical tolerances that do not have the symbol Ⓕ in the tolerance frame apply in the restrained (assembled) condition.

Drawings of non-rigid parts shall include the following:

- "ISO 10579-NR" should be indicated on the drawing in or near the title block.
- The Ⓕ symbol should be placed in the tolerance frame of all the geometrical tolerances where the geometric variations allowed are in the free state.
- The conditions in which the free state tolerances apply should be stated on the drawing . Examples are the "direction of gravity" or "orientation of the part" i.e. when the part is in a horizontal position.
- The conditions in which the part should be restrained to meet the drawing requirements.

An example showing tolerancing of a non-rigid (flexible) part according to ISO 10579 is given in example 1.

Example 1: Tolerancing of a non-rigid (flexible) part according to ISO 10579

ISO 10579 - NR Restrained condition:
Surface B to be mounted against a flat surface with 12 x M20 bolts torqued to
12 to 16 Nm while restraining datum feature C to the specified size limit.

Note:
The dimension $\varnothing 200 {\,}^{0}_{-1}$ in example 1 above is indicated as "average".
This size is calculated by taking the average of at least four measurements.

When a geometrical tolerance or datum reference is applied to a screw thread, the tolerance or datum is applied to the axis of thread derived from the PITCH cylinder, unless otherwise stated.

When a geometrical tolerance is applied to a spline or gear, the specific feature to which they apply i.e. pitch diameter PD, major diameter MD or minor (least) diameter LD must be stated beneath the tolerance frame, or beneath or adjacent to the datum frame.

Example 1: Geometrical tolerance applied to a screw thread

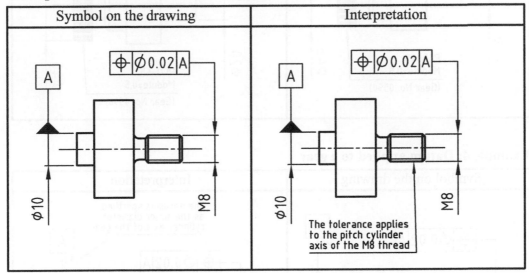

Example 2: Datum applied to the screw thread

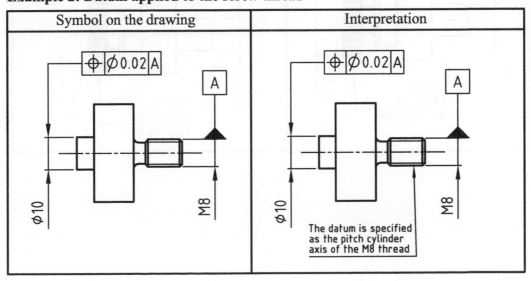

Example 3: Geometrical tolerance applied to a gear

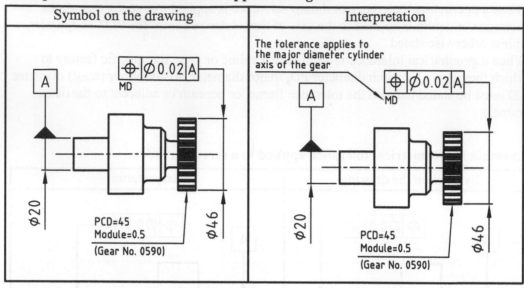

Symbol on the drawing	Interpretation

Example 4: Datum applied to a gear

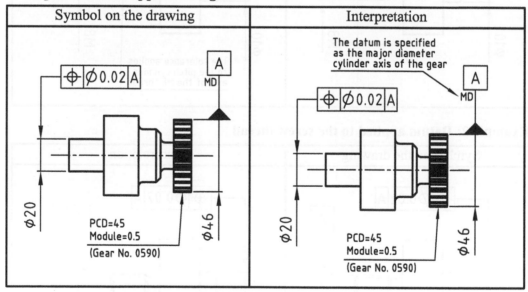

Symbol on the drawing	Interpretation

The **Principle of Independency** means that:
Each requirement for dimensional or geometrical tolerancing specified on a drawing shall be met INDEPENDENTLY unless a particular relationship is specified i.e.

Ⓜ Maximum material requirement, Ⓛ Least material requirement or
Ⓔ Envelope requirement.

If the Principle of Independency applies then it must be stated in the title block on the drawing as follows:

Tolerancing ISO 8015

Example 1:

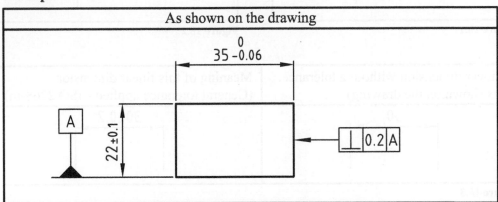

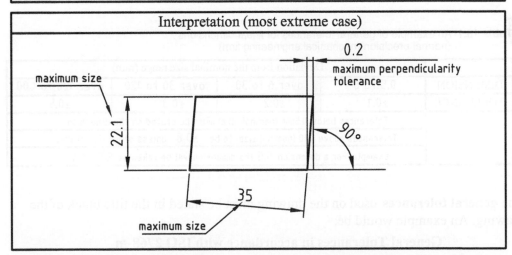

The linear size tolerances $35_{-0.06}^{\ 0}$ and 22 ± 0.1 are independent of the geometrical tolerance ⊥0.2A so they should be treated separately, independent of each other.

The Principle of Independency applies when using International Standards (ISO-standards) but not when using American National Standards (ANSI).

GENERAL TOLERANCES 18

General tolerances are tolerances that are to customary workshop accuracy, tolerances that a workshop can easily obtain. When tolerances are not stated on a dimension on a drawing then general tolerances apply to that dimension.

Each workshop will have its own standard general tolerances that it works to. It will have general tolerances for linear dimensions (see table 18.1) and general tolerances for geometrical tolerances (see table 18.2).

Linear dimension without a tolerance. (General tolerance applies)	Linear dimension with a tolerance. (General tolerance does not apply)
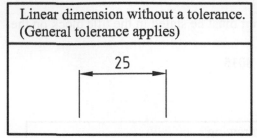	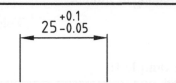

Figure 18.1   **Figure 18.2**

Linear dimension without a tolerance. (as shown on the drawing)	Meaning of this linear dimension (General tolerance applied - ISO 2768-m)
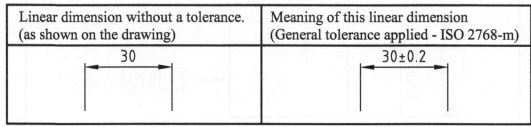	

Figure 18.3

Table 18.1 An example of general tolerances for linear dimensions
(normal precision mechanical engineering firm)

	Deviation from the nominal size range (mm)			
DIMENSION	0.5 to 6	over 6 to 30	over 30 to 120	over 120 to 400
TOLERANCE	±0.1	±0.2	±0.3	±0.5
	Tolerances below 0.5mm (nominal size) must be stated on the dimension			
	Tolerances above 400 (nominal size) to be ±0.8 unless otherwise stated			
	Example: For a dimension 5.5 the dimension will be taken as 5.5 ±0.1			

The **general tolerances** used on the drawing must be stated in the title block of the drawing. An example would be:

General Tolerances in accordance with ISO 2768-m

The general tolerances will be known by the workshop, so when a dimension on the drawing does not have a tolerance applied to it, the workshop will automatically apply the general tolerance to that dimension so as to know the tolerance limits before machining the part.

There are also **general tolerances for geometrical constraints.** When no geometrical tolerances are shown on the drawing then general geometrical tolerances will apply (see figure 18.4) if stated in the title block of the drawing. An example would be:

General tolerances in accordance with ISO 2768-K

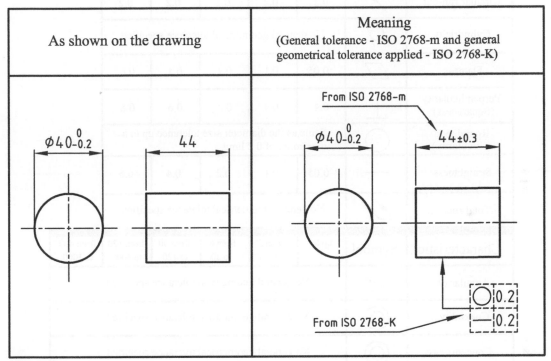

Figure 18.4

** Note: If general tolerances apply (see page 60) to your drawing then these values are not visibly shown on the drawing.

Table 18.2 An example of general geometrical tolerance values
(normal precision mechanical engineering firm)

Characteristic	Symbol	up to 10	from 10 to 30	from 30 to 100	from 100 to 300	from 300 to 1000	
Circular run-out	⟋	0.2	0.2	0.2	0.2	0.2	
Cylindricity	⌭	No general geometrical tolerance specified					
Flatness	▱	0.05	0.1	0.2	0.4	0.6	
Perpendicularity (Squareness)	⊥	0.4	0.4	0.4	0.6	0.8	
Roundness (Circularity)	◯	Same as the diameter size tolerance up to a maximum of 0.2 limit					
Straightness	—	0.05	0.1	0.2	0.4	0.6	
Total run-out	⟋⟋	No general geometrical tolerance specified					
Characteristic	Symbol	up to 3	from 3 to 6	from 6 to 30	from 30 to 120	from 120 to 400	from 400 to 1000
Angularity	∠	No general geometrical tolerance specified					
Coaxiality	◎	No general geometrical tolerance specified					
Concentricity	◎	No general geometrical tolerance specified					
Parallelism	//	No general geometrical tolerance specified					
Position	⊕	No general geometrical tolerance specified					
Profile of a line	⌒	No general geometrical tolerance specified					
Profile of a surface	⌓	No general geometrical tolerance specified					
Symmetry	⌯	0.6	0.6	0.6	0.6	0.8	1

If a better geometrical tolerance is needed, higher than the general level of table 18.2 then the geometrical tolerance must be quoted on the drawing in the tolerance frame.

An example of geometrical tolerance values (not general values) is given in table 18.3

** Note: These values are to be quoted on the drawing in the required tolerance frame.

Table 18.3 An example of geometrical tolerance values
(normal precision mechanical engineering firm)

Characteristic	Symbol	up to 10	from 10 to 30	from 30 to 100	from 100 to 300	from 300 to 1000	
Angularity	∠	0.02	0.06	0.25	0.8	2	
Circular run-out	↗	0.05	0.05	0.05	0.05	0.05	
Coaxiality	◎	0.05	0.05	0.1	0.15	0.2	
Concentricity	◎	0.05	0.05	0.1	0.15	0.2	
Cylindricity	⌀	0.25	0.25	0.25	0.3	0.4	
Flatness	▱	0.01	0.02	0.05	0.15	0.3	
Parallelism	//	0.01	0.03	0.1	0.3	1	
Perpendicularity (Squareness)	⊥	0.01	0.03	0.1	0.2	0.5	
Position	⊕	0.05	0.05	0.1	0.15	0.2	
Profile of a line	⌒	0.05	0.05	0.1	0.15	0.2	
Profile of a surface	◠	0.05	0.05	0.1	0.15	0.2	
Roundness (Circularity)	○	0.1	0.1	0.1	0.1	0.1	
Straightness	—	0.005	0.015	0.05	0.15	0.3	
Total run-out	↗↗	0.05	0.05	0.05	0.05	0.05	
Characteristic	Symbol	up to 3	from 3 to 6	from 6 to 30	from 30 to 120	from 120 to 400	from 400 to 1000
Symmetry	＝	0.05	0.1	0.2	0.3	0.5	0.8

Angular dimensions

Any angular dimensions that do not have individual tolerances indicated i.e.
should be classed as general angular dimensions and therefore **general angular
tolerances** will apply to these dimensions.

General tolerances for angular dimensions are usually specified in angular units and
not in linear units.

Table 18.4 An example of general tolerances for angular dimensions
(normal precision mechanical engineering firm)

	Deviation from the nominal size range (mm)			
DIMENSION	up to 10	over 10 to 50	over 50 to 120	over 120 to 400
TOLERANCE	±1°	±30'	±20'	±10'
	Tolerances above 400 (nominal size) to be ±5' unless otherwise stated			
	Example: For a dimension 6° the dimension will be taken as 6° ±1°			

General tolerances used on a drawing for angular dimensions must be stated in the title
block. If ISO 2768 is the general tolerance then the following note should be stated in
the title block of the drawing:

General tolerances in accordance with ISO 2768-m

This note covers general tolerances for both linear and angular dimensions (the
tolerance class in this case is **m** for medium class)

Note:
ISO 2768 for linear and angular dimensions has four tolerance classes, fine (f),
medium (m), coarse (c) and very coarse (v).

PART 3

Geometrical tolerancing examples

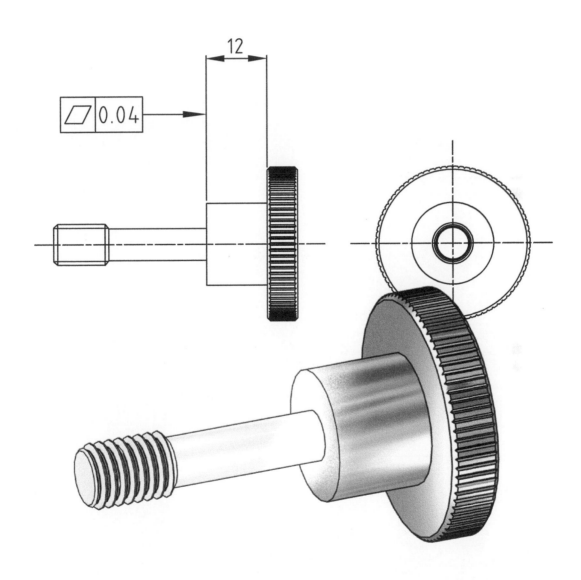

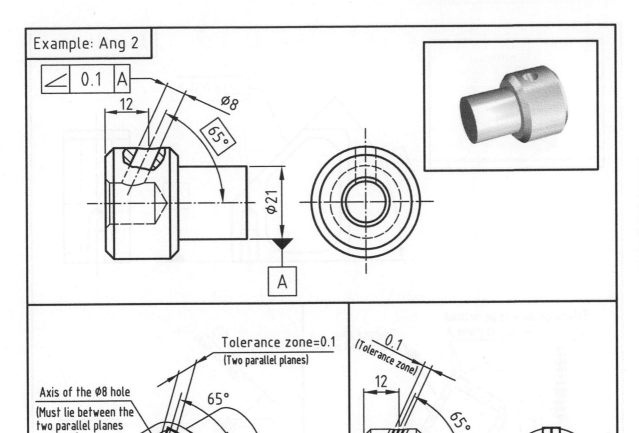

Example: Ang 2

| ∠ | 0.1 | A |

12   Ø8   65°   Ø21

A

Tolerance zone=0.1
(Two parallel planes)

Axis of the Ø8 hole
(Must lie between the
two parallel planes
0.1 apart)

65°

Datum axis A
(Feature Ø21)

Tolerance zone to be inclined
at 65° to datum axis

0.1
(Tolerance zone)

12   65°

ANGULARITY

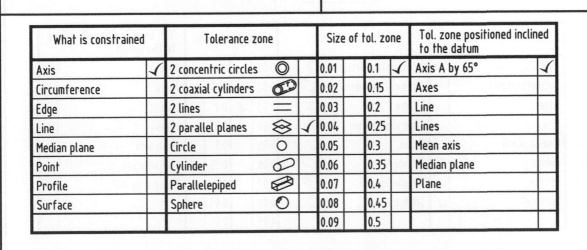

What is constrained		Tolerance zone		Size of tol. zone		Tol. zone positioned inclined to the datum	
Axis	✓	2 concentric circles ◎		0.01	0.1 ✓	Axis A by 65°	✓
Circumference		2 coaxial cylinders		0.02	0.15	Axes	
Edge		2 lines =		0.03	0.2	Line	
Line		2 parallel planes ⬙	✓	0.04	0.25	Lines	
Median plane		Circle ○		0.05	0.3	Mean axis	
Point		Cylinder		0.06	0.35	Median plane	
Profile		Parallelepiped		0.07	0.4	Plane	
Surface		Sphere ○		0.08	0.45		
				0.09	0.5		

ANGULARITY

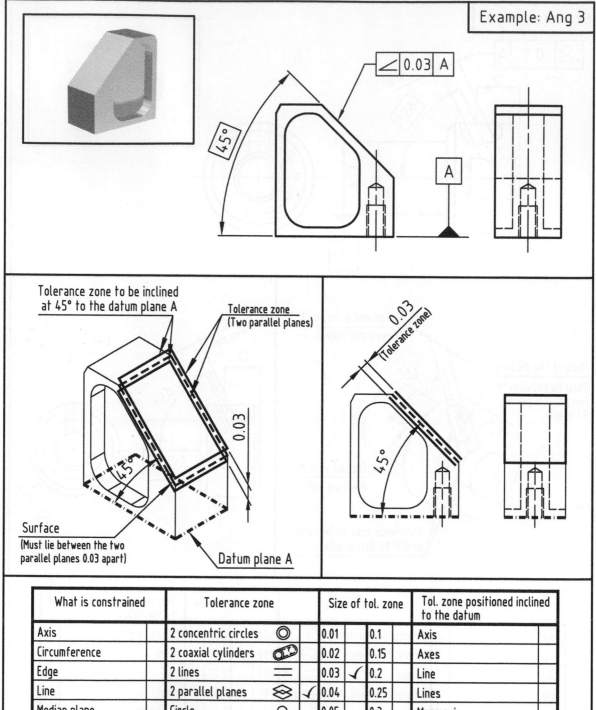

Tolerance zone to be inclined at 45° to the datum plane A

Tolerance zone (Two parallel planes)

0.03

45°

Surface
(Must lie between the two parallel planes 0.03 apart)

Datum plane A

0.03 (Tolerance zone)

45°

What is constrained		Tolerance zone		Size of tol. zone				Tol. zone positioned inclined to the datum	
Axis		2 concentric circles	◎	0.01		0.1		Axis	
Circumference		2 coaxial cylinders	⬯	0.02		0.15		Axes	
Edge		2 lines	=	0.03	✓	0.2		Line	
Line		2 parallel planes	⬦	✓	0.04	0.25		Lines	
Median plane		Circle	○	0.05		0.3		Mean axis	
Point		Cylinder	⬭	0.06		0.35		Median plane	
Profile		Parallelepiped	▱	0.07		0.4		Plane A by 45°	✓
Surface	✓	Sphere	◓	0.08		0.45			
				0.09		0.5			

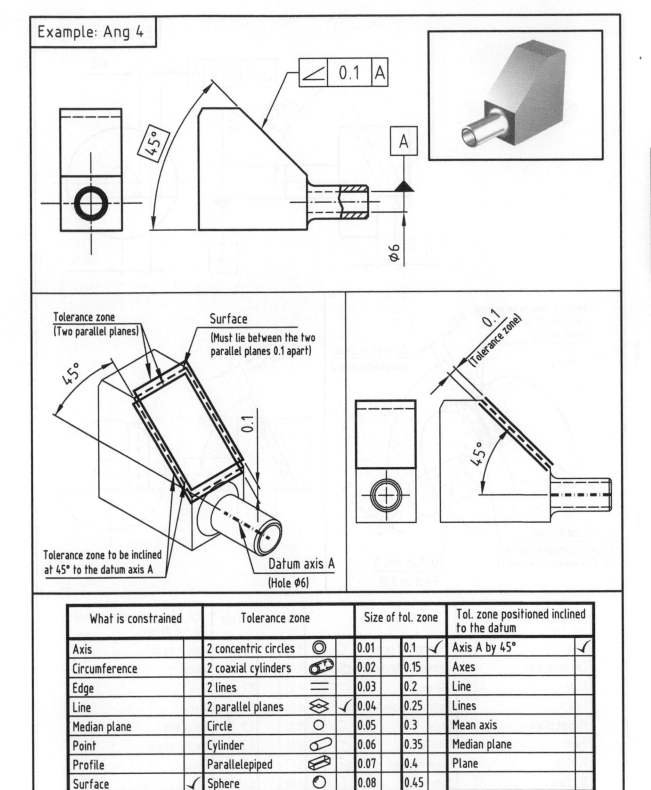

Example: Ang 4

∠ 0.1 A

45°

A

Ø6

ANGULARITY

Tolerance zone
(Two parallel planes)

Surface
(Must lie between the two
parallel planes 0.1 apart)

45°

0.1

Tolerance zone to be inclined
at 45° to the datum axis A

Datum axis A
(Hole Ø6)

0.1
(Tolerance zone)

45°

What is constrained		Tolerance zone		Size of tol. zone				Tol. zone positioned inclined to the datum	
Axis		2 concentric circles	◎	0.01		0.1	✓	Axis A by 45°	✓
Circumference		2 coaxial cylinders	⬭	0.02		0.15		Axes	
Edge		2 lines	=	0.03		0.2		Line	
Line		2 parallel planes	⬖	✓	0.04	0.25		Lines	
Median plane		Circle	○	0.05		0.3		Mean axis	
Point		Cylinder	⬭	0.06		0.35		Median plane	
Profile		Parallelepiped	⬭	0.07		0.4		Plane	
Surface	✓	Sphere	⊘	0.08		0.45			
				0.09		0.5			

ANGULARITY

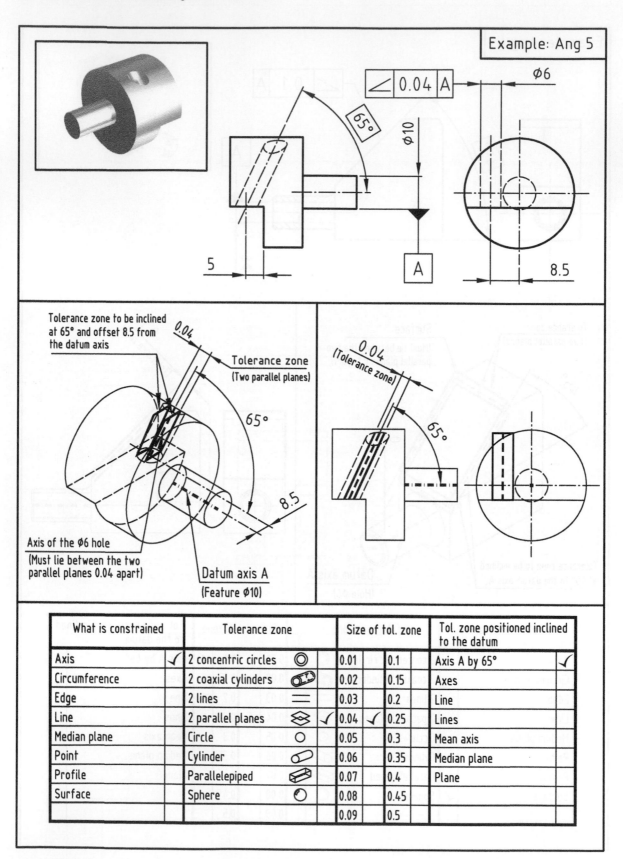

Example: Ang 5

∠ | 0.04 | A

Ø6

Ø10

65°

A

5

A

8.5

Tolerance zone to be inclined at 65° and offset 8.5 from the datum axis

0.04

Tolerance zone
(Two parallel planes)

65°

8.5

Axis of the Ø6 hole
(Must lie between the two parallel planes 0.04 apart)

Datum axis A
(Feature Ø10)

0.04
(Tolerance zone)

65°

What is constrained		Tolerance zone		Size of tol. zone				Tol. zone positioned inclined to the datum	
Axis	✓	2 concentric circles	◎	0.01		0.1		Axis A by 65°	✓
Circumference		2 coaxial cylinders	⌽	0.02		0.15		Axes	
Edge		2 lines	═	0.03		0.2		Line	
Line		2 parallel planes	⬨ ✓	0.04	✓	0.25		Lines	
Median plane		Circle	○	0.05		0.3		Mean axis	
Point		Cylinder	�－	0.06		0.35		Median plane	
Profile		Parallelepiped	⬗	0.07		0.4		Plane	
Surface		Sphere	◓	0.08		0.45			
				0.09		0.5			

Example: Ang 6

∠ | 0.1 | A–B

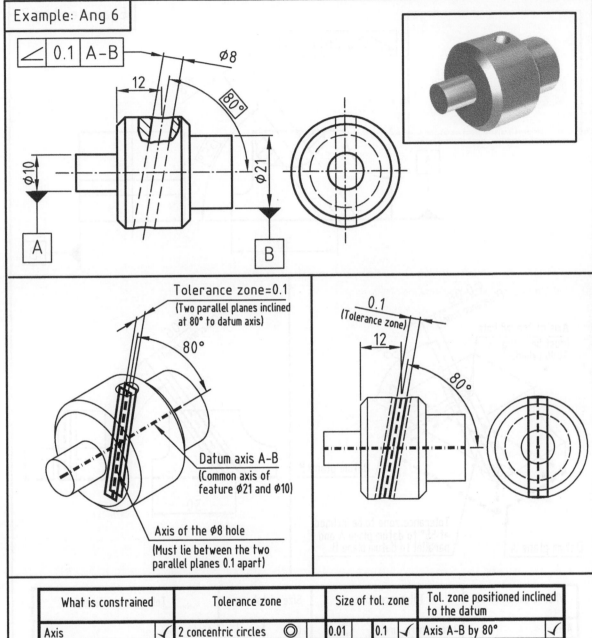

ANGULARITY

Tolerance zone = 0.1
(Two parallel planes inclined at 80° to datum axis)

80°

Datum axis A–B
(Common axis of feature ⌀21 and ⌀10)

Axis of the ⌀8 hole
(Must lie between the two parallel planes 0.1 apart)

0.1
(Tolerance zone)

12

80°

What is constrained		Tolerance zone		Size of tol. zone		Tol. zone positioned inclined to the datum	
Axis	✓	2 concentric circles ◎		0.01	0.1 ✓	Axis A–B by 80°	✓
Circumference		2 coaxial cylinders		0.02	0.15	Axes	
Edge		2 lines =		0.03	0.2	Line	
Line		2 parallel planes ⬨	✓	0.04	0.25	Lines	
Median plane		Circle ○		0.05	0.3	Mean axis	
Point		Cylinder ⌀		0.06	0.35	Median plane	
Profile		Parallelepiped ▱		0.07	0.4	Plane	
Surface		Sphere ◯		0.08	0.45		
				0.09	0.5		

ANGULARITY

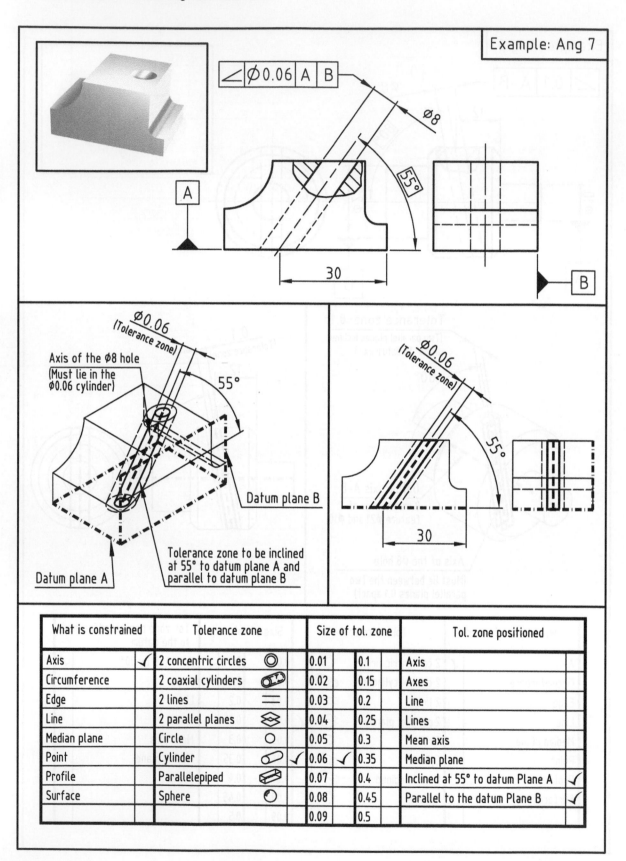

Example: Ang 7

∠ | ⌀0.06 | A | B

⌀8

55°

A

30

B

⌀0.06
(Tolerance zone)

Axis of the ⌀8 hole
(Must lie in the
⌀0.06 cylinder)

55°

Datum plane B

Tolerance zone to be inclined
at 55° to datum plane A and
parallel to datum plane B

Datum plane A

⌀0.06
(Tolerance zone)

55°

30

What is constrained		Tolerance zone		Size of tol. zone			Tol. zone positioned	
Axis	✓	2 concentric circles	◎	0.01		0.1	Axis	
Circumference		2 coaxial cylinders	⌀	0.02		0.15	Axes	
Edge		2 lines	═	0.03		0.2	Line	
Line		2 parallel planes	⬗	0.04		0.25	Lines	
Median plane		Circle	○	0.05		0.3	Mean axis	
Point		Cylinder	⌀ ✓	0.06 ✓		0.35	Median plane	
Profile		Parallelepiped	▱	0.07		0.4	Inclined at 55° to datum Plane A	✓
Surface		Sphere	◉	0.08		0.45	Parallel to the datum Plane B	✓
				0.09		0.5		

Example: Cro 1

| ⟋ | 0.08 | A-B |

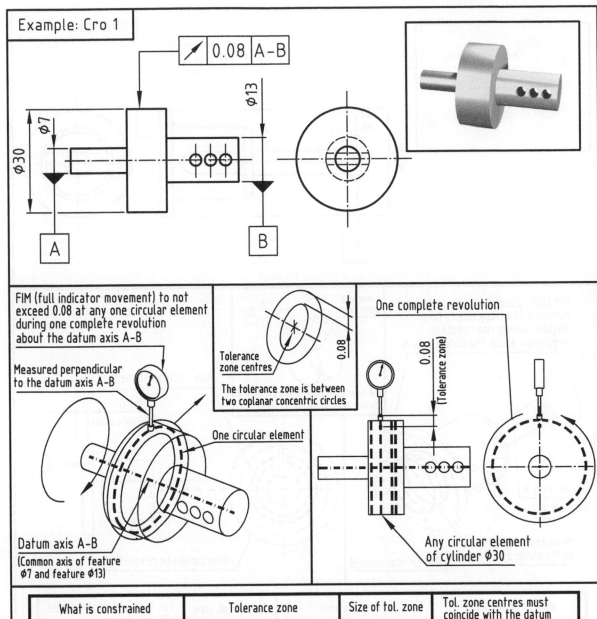

FIM (full indicator movement) to not exceed 0.08 at any one circular element during one complete revolution about the datum axis A-B

Measured perpendicular to the datum axis A-B

One circular element

Datum axis A-B
(Common axis of feature ⌀7 and feature ⌀13)

Tolerance zone centres

The tolerance zone is between two coplanar concentric circles

One complete revolution

0.08 (Tolerance zone)

Any circular element of cylinder ⌀30

What is constrained		Tolerance zone		Size of tol. zone		Tol. zone centres must coincide with the datum	
Axis		2 concentric circles	◎ ✓	0.01	0.1	Axis A-B	✓
Circumference		2 coaxial cylinders	⌀	0.02	0.15	Axes	
Edge		2 lines	=	0.03	0.2	Line	
Line		2 parallel planes	⬦	0.04	0.25	Lines	
Median plane		Circle	○	0.05	0.3	Mean axis	
Point		Cylinder	⌀	0.06	0.35	Median plane	
Profile		Parallelepiped	▱	0.07	0.4	Plane	
Surface		Sphere	○	0.08 ✓	0.45		
Circular element of a surface	✓			0.09	0.5		

CIRCULAR RUN-OUT

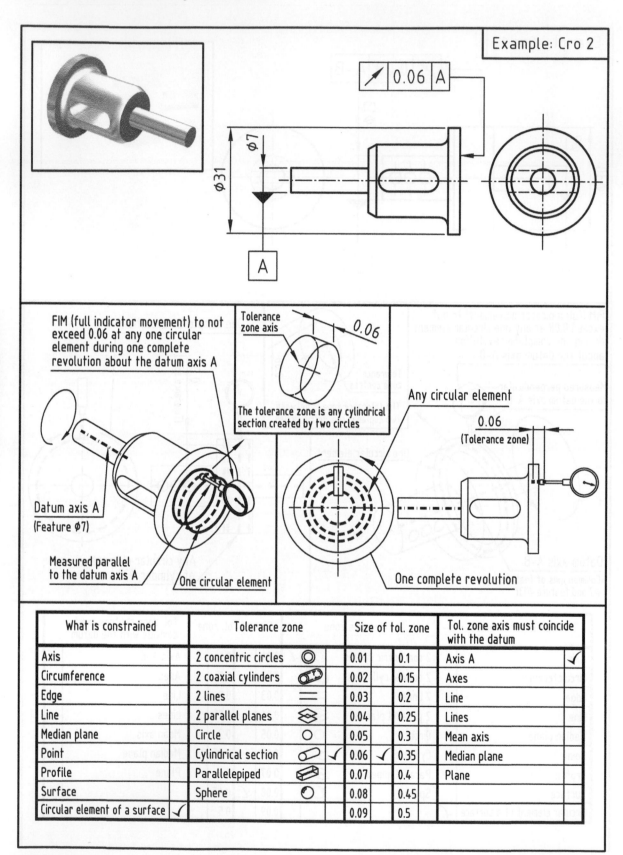

Example: Cro 2

FIM (full indicator movement) to not exceed 0.06 at any one circular element during one complete revolution about the datum axis A

Tolerance zone axis

The tolerance zone is any cylindrical section created by two circles

Datum axis A (Feature ⌀7)

Measured parallel to the datum axis A

One circular element

Any circular element

0.06 (Tolerance zone)

One complete revolution

What is constrained	Tolerance zone		Size of tol. zone		Tol. zone axis must coincide with the datum	
Axis	2 concentric circles	◎	0.01	0.1	Axis A	✓
Circumference	2 coaxial cylinders	⌀	0.02	0.15	Axes	
Edge	2 lines	═	0.03	0.2	Line	
Line	2 parallel planes	⬦	0.04	0.25	Lines	
Median plane	Circle	○	0.05	0.3	Mean axis	
Point	Cylindrical section	⌀ ✓	0.06 ✓	0.35	Median plane	
Profile	Parallelepiped	▱	0.07	0.4	Plane	
Surface	Sphere	◓	0.08	0.45		
Circular element of a surface ✓			0.09	0.5		

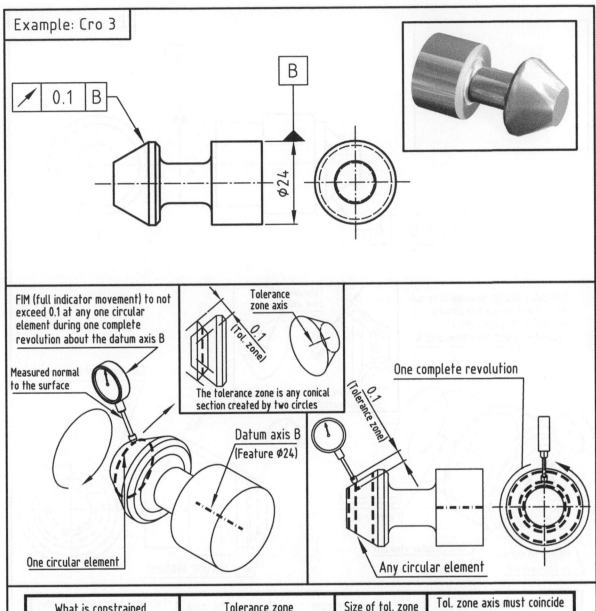

Example: Cro 3

B

⟋ 0.1 B

⌀24

B (Datum)

FIM (full indicator movement) to not exceed 0.1 at any one circular element during one complete revolution about the datum axis B

Measured normal to the surface

One circular element

Tolerance zone axis

0.1 (Tol. zone)

The tolerance zone is any conical section created by two circles

Datum axis B (Feature ⌀24)

One complete revolution

0.1 (Tolerance zone)

Any circular element

What is constrained		Tolerance zone		Size of tol. zone			Tol. zone axis must coincide with the datum	
Axis		2 concentric circles	◎	0.01	0.1	✓	Axis B	✓
Circumference		2 coaxial cylinders	⌀⌀	0.02	0.15		Axes	
Edge		2 lines	=	0.03	0.2		Line	
Line		2 parallel planes	⬨	0.04	0.25		Lines	
Median plane		Circle	○	0.05	0.3		Mean axis	
Point		Cylinder	⌀	0.06	0.35		Median plane	
Profile		Parallelepiped	▱	0.07	0.4		Plane	
Surface		Sphere	◓	0.08	0.45			
Circular element of a surface	✓	Conical section	◗	✓	0.09	0.5		

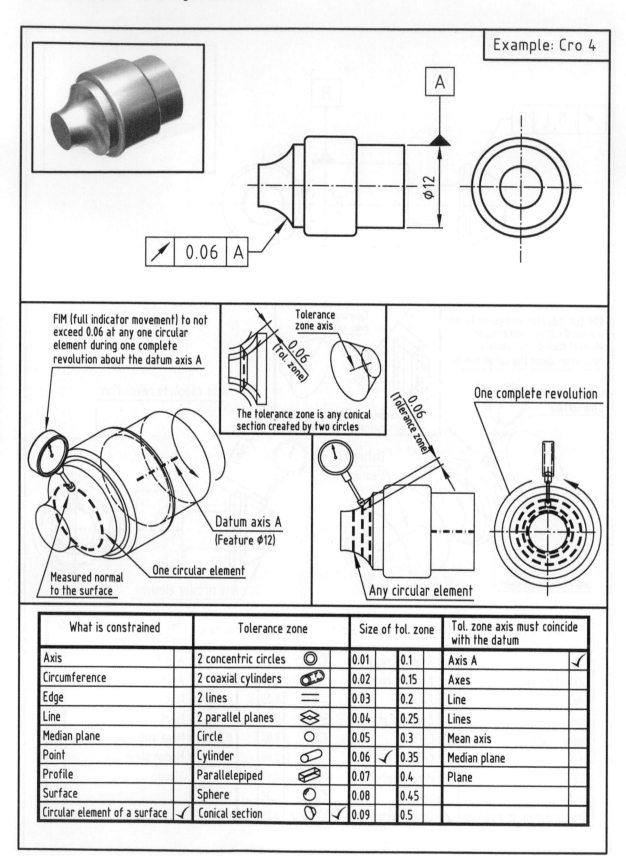

Example: Cro 4

A

Ø12

⌿ | 0.06 | A

CIRCULAR RUN-OUT

FIM (full indicator movement) to not exceed 0.06 at any one circular element during one complete revolution about the datum axis A

Tolerance zone axis

0.06 (Tol. zone)

The tolerance zone is any conical section created by two circles

One complete revolution

0.06 (Tolerance zone)

Datum axis A (Feature Ø12)

One circular element

Measured normal to the surface

Any circular element

What is constrained		Tolerance zone		Size of tol. zone		Tol. zone axis must coincide with the datum	
Axis		2 concentric circles	◎	0.01	0.1	Axis A	✓
Circumference		2 coaxial cylinders	⬭	0.02	0.15	Axes	
Edge		2 lines	═	0.03	0.2	Line	
Line		2 parallel planes	❖	0.04	0.25	Lines	
Median plane		Circle	○	0.05	0.3	Mean axis	
Point		Cylinder	⬭	0.06 ✓	0.35	Median plane	
Profile		Parallelepiped	▱	0.07	0.4	Plane	
Surface		Sphere	◓	0.08	0.45		
Circular element of a surface	✓	Conical section	◗ ✓	0.09	0.5		

Example: Cro 5

A

⌀9

↗ 0.05 A B

B

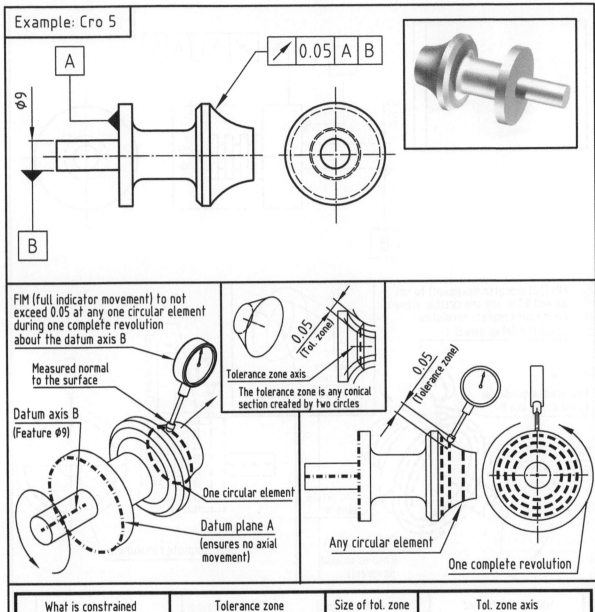

FIM (full indicator movement) to not exceed 0.05 at any one circular element during one complete revolution about the datum axis B

Measured normal to the surface

Datum axis B (Feature ⌀9)

One circular element

Datum plane A (ensures no axial movement)

0.05 (Tol. zone)

Tolerance zone axis

The tolerance zone is any conical section created by two circles

0.05 (Tolerance zone)

Any circular element

One complete revolution

What is constrained		Tolerance zone		Size of tol. zone				Tol. zone axis	
Axis		2 concentric circles	◎	0.01		0.1		Must coincide with datum axis B	✓
Circumference		2 coaxial cylinders	⌀	0.02		0.15		Axes	
Edge		2 lines	═	0.03		0.2		Line	
Line		2 parallel planes	◈	0.04		0.25		Lines	
Median plane		Circle	○	0.05	✓	0.3		Mean axis	
Point		Cylinder	⌀	0.06		0.35		Median plane	
Profile		Parallelepiped	▱	0.07		0.4		To be at 90° to datum plane A	✓
Surface		Sphere	○	0.08		0.45			
Circular element of a surface	✓	Conical section	◑	✓ 0.09		0.5			

CIRCULAR RUN-OUT

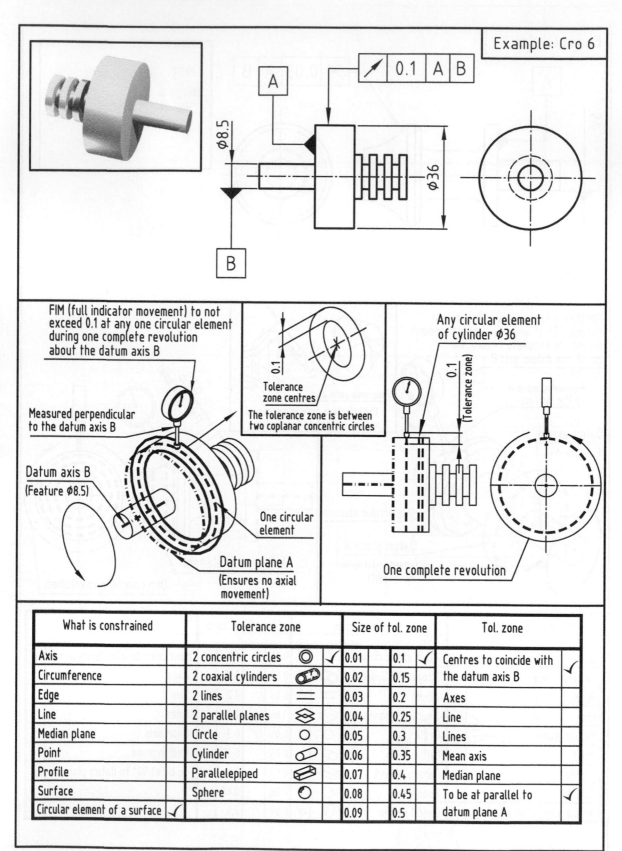

Example: Cro 6

⟋ | 0.1 | A | B

Ø8.5

Ø36

FIM (full indicator movement) to not exceed 0.1 at any one circular element during one complete revolution about the datum axis B

Measured perpendicular to the datum axis B

Datum axis B
(Feature Ø8.5)

Tolerance zone centres

The tolerance zone is between two coplanar concentric circles

One circular element

Datum plane A
(Ensures no axial movement)

Any circular element of cylinder Ø36

0.1 (Tolerance zone)

One complete revolution

What is constrained	Tolerance zone		Size of tol. zone			Tol. zone	
Axis	2 concentric circles	◎ ✓	0.01	0.1	✓	Centres to coincide with	✓
Circumference	2 coaxial cylinders	⬭	0.02	0.15		the datum axis B	
Edge	2 lines	═	0.03	0.2		Axes	
Line	2 parallel planes	⬙	0.04	0.25		Line	
Median plane	Circle	○	0.05	0.3		Lines	
Point	Cylinder	⬭	0.06	0.35		Mean axis	
Profile	Parallelepiped	▱	0.07	0.4		Median plane	
Surface	Sphere	◓	0.08	0.45		To be at parallel to	✓
Circular element of a surface	✓		0.09	0.5		datum plane A	

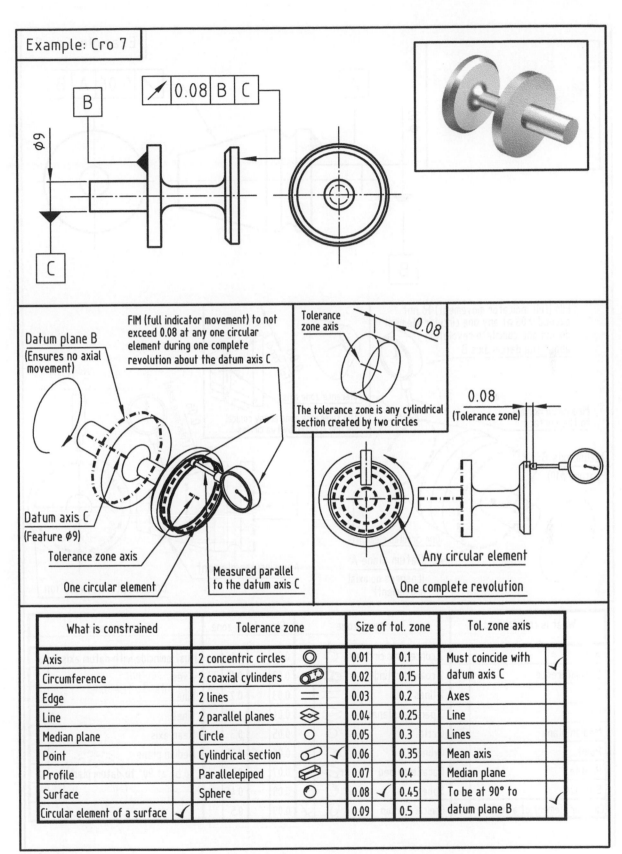

Example: Cro 7

B

Ø9

↗ 0.08 B C

C

FIM (full indicator movement) to not exceed 0.08 at any one circular element during one complete revolution about the datum axis C

Datum plane B
(Ensures no axial movement)

Tolerance zone axis

0.08

The tolerance zone is any cylindrical section created by two circles

0.08
(Tolerance zone)

Datum axis C
(Feature Ø9)

Tolerance zone axis

One circular element

Measured parallel to the datum axis C

Any circular element

One complete revolution

What is constrained	Tolerance zone		Size of tol. zone		Tol. zone axis	
Axis	2 concentric circles	◎	0.01	0.1	Must coincide with	✓
Circumference	2 coaxial cylinders	⌼	0.02	0.15	datum axis C	
Edge	2 lines	=	0.03	0.2	Axes	
Line	2 parallel planes	⬦	0.04	0.25	Line	
Median plane	Circle	○	0.05	0.3	Lines	
Point	Cylindrical section	⌼ ✓	0.06	0.35	Mean axis	
Profile	Parallelepiped	▱	0.07	0.4	Median plane	
Surface	Sphere	○	0.08 ✓	0.45	To be at 90° to	
Circular element of a surface ✓			0.09	0.5	datum plane B	✓

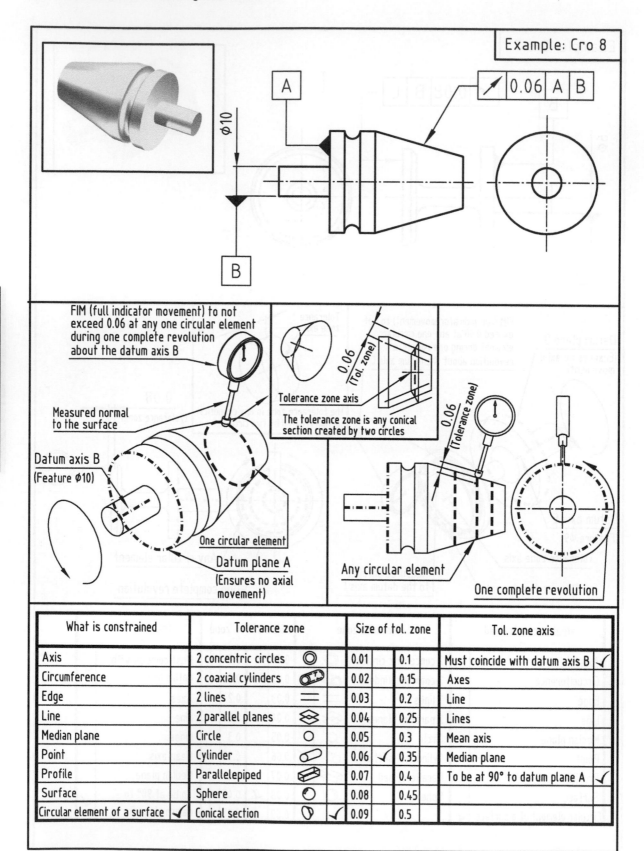

Example: Cro 8

A

B

⌀10

∕ 0.06 A B

CIRCULAR RUN-OUT

FIM (full indicator movement) to not exceed 0.06 at any one circular element during one complete revolution about the datum axis B

Measured normal to the surface

Datum axis B
(Feature ⌀10)

One circular element

Datum plane A
(Ensures no axial movement)

0.06
(Tol. zone)

Tolerance zone axis

The tolerance zone is any conical section created by two circles

0.06
(Tolerance zone)

Any circular element

One complete revolution

What is constrained		Tolerance zone		Size of tol. zone			Tol. zone axis	
Axis		2 concentric circles	◎	0.01		0.1	Must coincide with datum axis B	✓
Circumference		2 coaxial cylinders		0.02		0.15	Axes	
Edge		2 lines	=	0.03		0.2	Line	
Line		2 parallel planes	⬥	0.04		0.25	Lines	
Median plane		Circle	○	0.05		0.3	Mean axis	
Point		Cylinder	⬭	0.06	✓	0.35	Median plane	
Profile		Parallelepiped	▱	0.07		0.4	To be at 90° to datum plane A	✓
Surface		Sphere	◗	0.08		0.45		
Circular element of a surface	✓	Conical section	◑	✓	0.09		0.5		

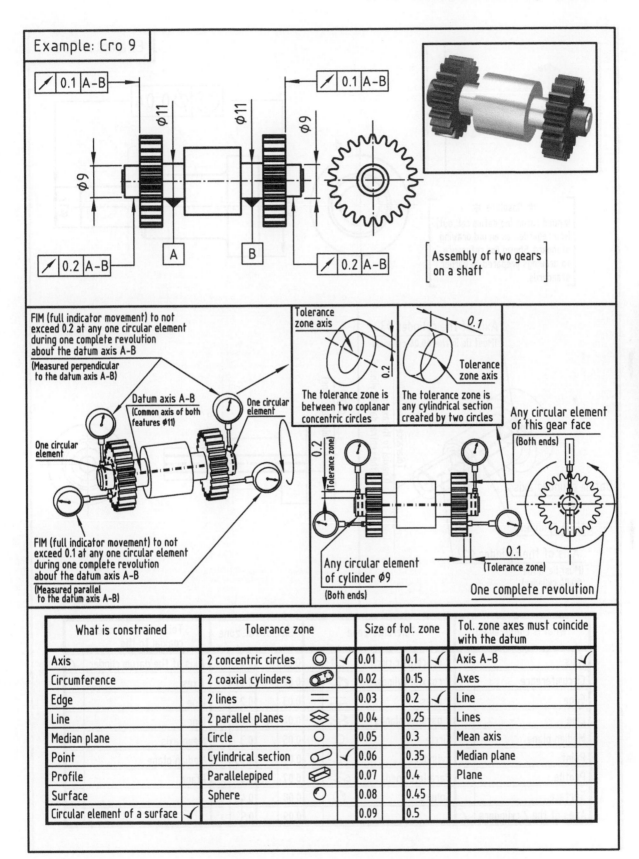

Example: Cro 9

| 0.1 | A–B |

Ø11   Ø11   Ø9

Ø9

| A | | B |

| 0.2 | A–B | | 0.2 | A–B |

Assembly of two gears on a shaft

FIM (full indicator movement) to not exceed 0.2 at any one circular element during one complete revolution about the datum axis A–B
(Measured perpendicular to the datum axis A–B)

Datum axis A–B
(Common axis of both features Ø11)

One circular element

One circular element

FIM (full indicator movement) to not exceed 0.1 at any one circular element during one complete revolution about the datum axis A–B
(Measured parallel to the datum axis A–B)

Tolerance zone axis

The tolerance zone is between two coplanar concentric circles

0.2

0.1

Tolerance zone axis

The tolerance zone is any cylindrical section created by two circles

0.2 (Tolerance zone)

Any circular element of cylinder Ø9
(Both ends)

Any circular element of this gear face
(Both ends)

0.1
(Tolerance zone)

One complete revolution

What is constrained		Tolerance zone			Size of tol. zone		Tol. zone axes must coincide with the datum	
Axis		2 concentric circles	◎	✓	0.01	0.1 ✓	Axis A–B	✓
Circumference		2 coaxial cylinders	⌀		0.02	0.15	Axes	
Edge		2 lines	=		0.03	0.2 ✓	Line	
Line		2 parallel planes	⬥		0.04	0.25	Lines	
Median plane		Circle	○		0.05	0.3	Mean axis	
Point		Cylindrical section	⌀	✓	0.06	0.35	Median plane	
Profile		Parallelepiped	▱		0.07	0.4	Plane	
Surface		Sphere	◯		0.08	0.45		
Circular element of a surface	✓				0.09	0.5		

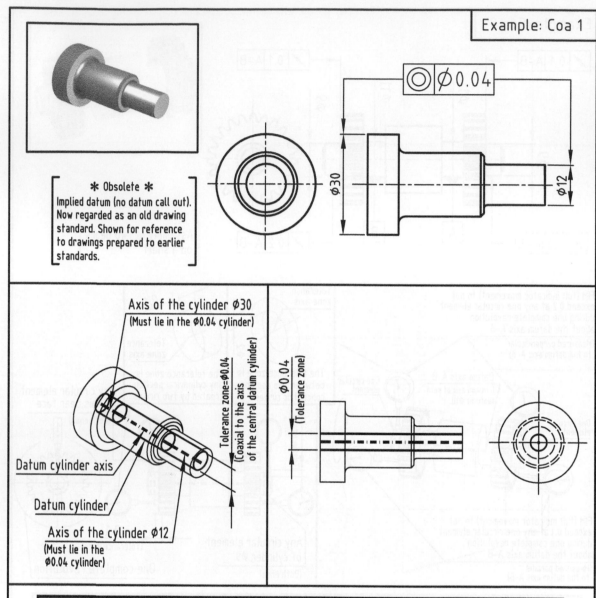

◎ | ⌀0.04

⌀30

⌀12

* Obsolete *
Implied datum (no datum call out).
Now regarded as an old drawing
standard. Shown for reference
to drawings prepared to earlier
standards.

Axis of the cylinder ⌀30
(Must lie in the ⌀0.04 cylinder)

Tolerance zone=⌀0.04
(Coaxial to the axis
of the central datum cylinder)

⌀0.04
(Tolerance zone)

Datum cylinder axis

Datum cylinder

Axis of the cylinder ⌀12
(Must lie in the
⌀0.04 cylinder)

COAXIALITY

What is constrained	Tolerance zone		Size of tol. zone		Tol. zone positioned coaxial to the	
Axis	2 concentric circles	◎	0.01	0.1	Axis of the datum clinder	✓
Circumference	2 coaxial cylinders	⬭	0.02	0.15	Axes	
Edge	2 lines	═	0.03	0.2	Line	
Line	2 parallel planes	⬦	0.04 ✓	0.25	Lines	
Median plane	Circle	○	0.05	0.3	Mean axis	
Point	Cylinder	⬭ ✓	0.06	0.35	Median plane	
Profile	Parallelepiped	⬠	0.07	0.4	Plane	
Surface	Sphere	⬯	0.08	0.45		
Axes of the 2 cylinders ✓			0.09	0.5		

Example: Coa 2

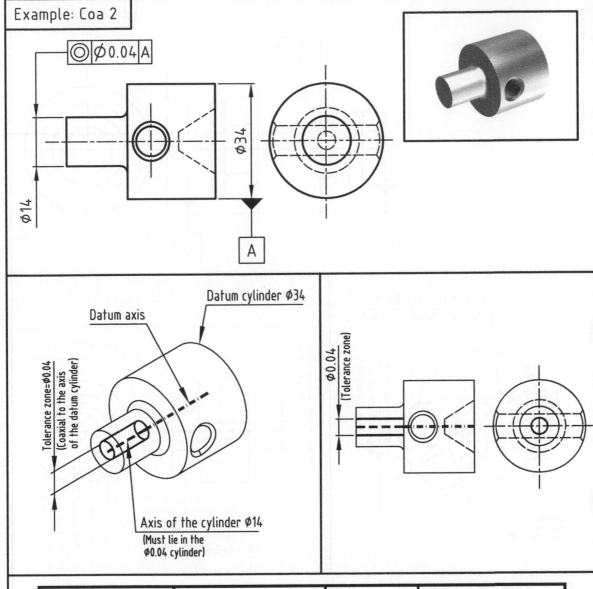

Datum cylinder ⌀34

Datum axis

Tolerance zone=⌀0.04
(Coaxial to the axis
of the datum cylinder)

Axis of the cylinder ⌀14
(Must lie in the
⌀0.04 cylinder)

⌀0.04
(Tolerance zone)

COAXIALITY

What is constrained		Tolerance zone		Size of tol. zone		Tol. zone positioned coaxial to the datum	
Axis	✓	2 concentric circles ◎		0.01	0.1	Axis	✓
Circumference		2 coaxial cylinders		0.02	0.15	Axes	
Edge		2 lines =		0.03	0.2	Line	
Line		2 parallel planes ⬨		0.04 ✓	0.25	Lines	
Median plane		Circle ○		0.05	0.3	Mean axis	
Point		Cylinder ⬭	✓	0.06	0.35	Median plane	
Profile		Parallelepiped ⬭		0.07	0.4	Plane	
Surface		Sphere ◯		0.08	0.45		
				0.09	0.5		

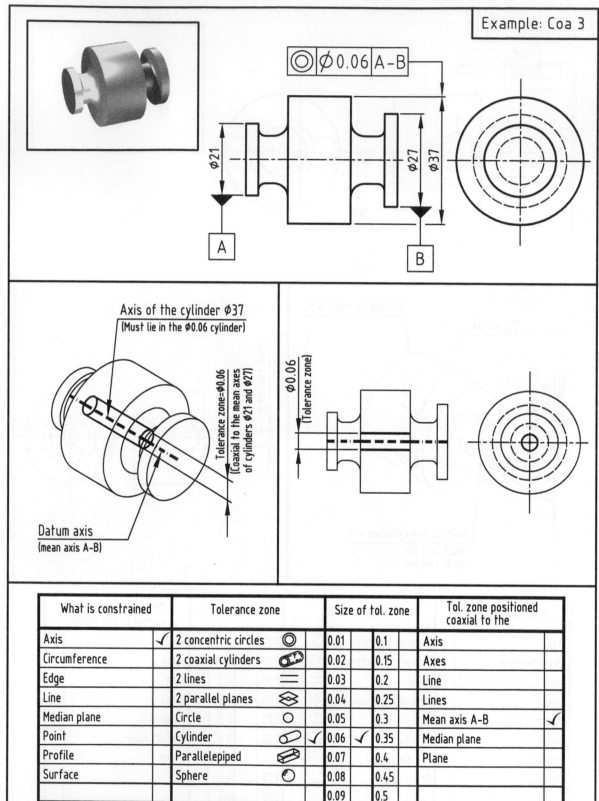

Example: Coa 3

◎ | ⌀0.06 | A–B

⌀21   ⌀27   ⌀37

A   B

Axis of the cylinder ⌀37
(Must lie in the ⌀0.06 cylinder)

Tolerance zone=⌀0.06
(Coaxial to the mean axes
of cylinders ⌀21 and ⌀27)

Datum axis
(mean axis A–B)

⌀0.06
(Tolerance zone)

COAXIALITY

What is constrained		Tolerance zone		Size of tol. zone		Tol. zone positioned coaxial to the	
Axis	✓	2 concentric circles	◎	0.01	0.1	Axis	
Circumference		2 coaxial cylinders	⌀	0.02	0.15	Axes	
Edge		2 lines	═	0.03	0.2	Line	
Line		2 parallel planes	⬨	0.04	0.25	Lines	
Median plane		Circle	○	0.05	0.3	Mean axis A–B	✓
Point		Cylinder	⌀	✓ 0.06	✓ 0.35	Median plane	
Profile		Parallelepiped	▱	0.07	0.4	Plane	
Surface		Sphere	◯	0.08	0.45		
				0.09	0.5		

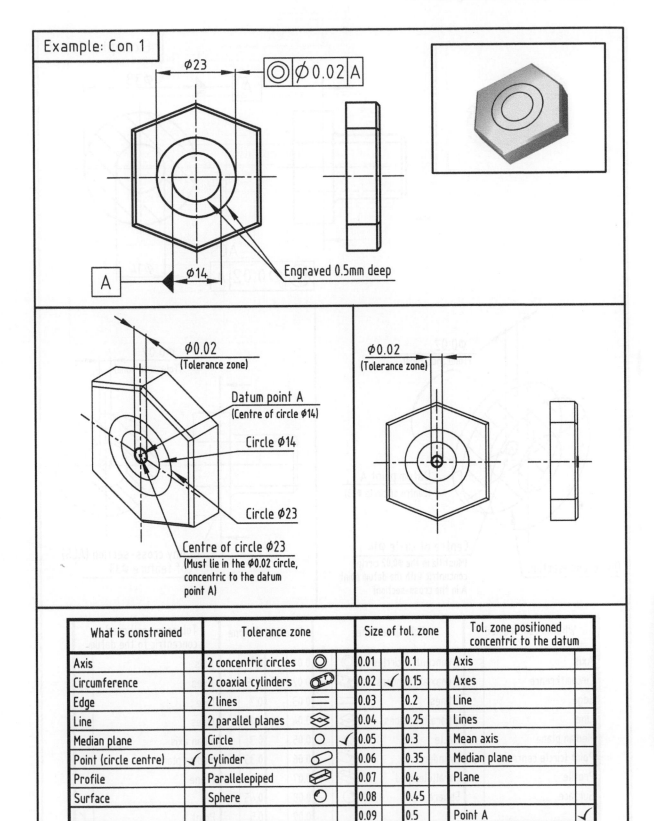

Example: Con 1

Ø23

◎ Ø0.02 A

Engraved 0.5mm deep

A

Ø14

Ø0.02
(Tolerance zone)

Datum point A
(Centre of circle Ø14)

Circle Ø14

Circle Ø23

Centre of circle Ø23
(Must lie in the Ø0.02 circle,
concentric to the datum
point A)

Ø0.02
(Tolerance zone)

What is constrained		Tolerance zone		Size of tol. zone		Tol. zone positioned concentric to the datum	
Axis		2 concentric circles ◎		0.01	0.1	Axis	
Circumference		2 coaxial cylinders		0.02 ✓	0.15	Axes	
Edge		2 lines		0.03	0.2	Line	
Line		2 parallel planes ⬦		0.04	0.25	Lines	
Median plane		Circle ○ ✓		0.05	0.3	Mean axis	
Point (circle centre) ✓		Cylinder ⬭		0.06	0.35	Median plane	
Profile		Parallelepiped ▱		0.07	0.4	Plane	
Surface		Sphere ⬮		0.08	0.45		
				0.09	0.5	Point A	✓

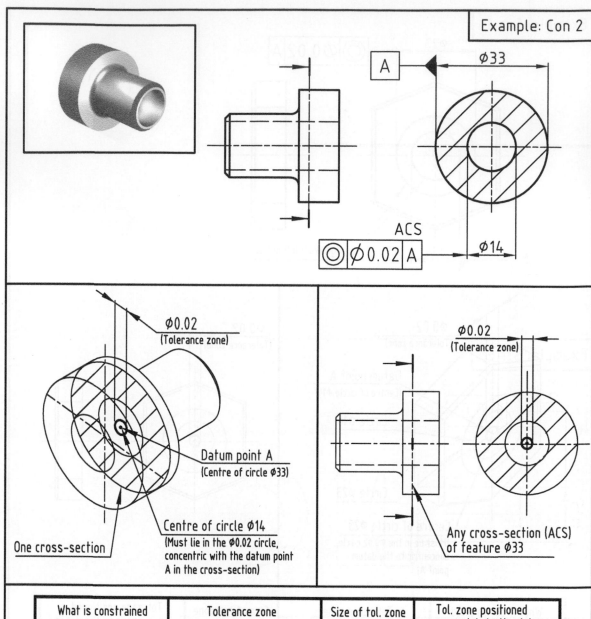

Example: Con 2

φ33

A

ACS

◎ | φ0.02 | A

φ14

φ0.02
(Tolerance zone)

Datum point A
(Centre of circle φ33)

One cross-section

Centre of circle φ14
(Must lie in the φ0.02 circle,
concentric with the datum point
A in the cross-section)

φ0.02
(Tolerance zone)

Any cross-section (ACS)
of feature φ33

CONCENTRICITY

What is constrained		Tolerance zone		Size of tol. zone		Tol. zone positioned concentric to the datum	
Axis		2 concentric circles	◎	0.01	0.1	Axis	
Circumference		2 coaxial cylinders	⌀	0.02 ✓	0.15	Axes	
Edge		2 lines	=	0.03	0.2	Line	
Line		2 parallel planes	⬦	0.04	0.25	Lines	
Median plane		Circle	○ ✓	0.05	0.3	Mean axis	
Point (circle centre)	✓	Cylinder	⌀	0.06	0.35	Median plane	
Profile		Parallelepiped	▱	0.07	0.4	Plane	
Surface		Sphere	◐	0.08	0.45		
				0.09	0.5	Point A	✓

Example: Cyl 1

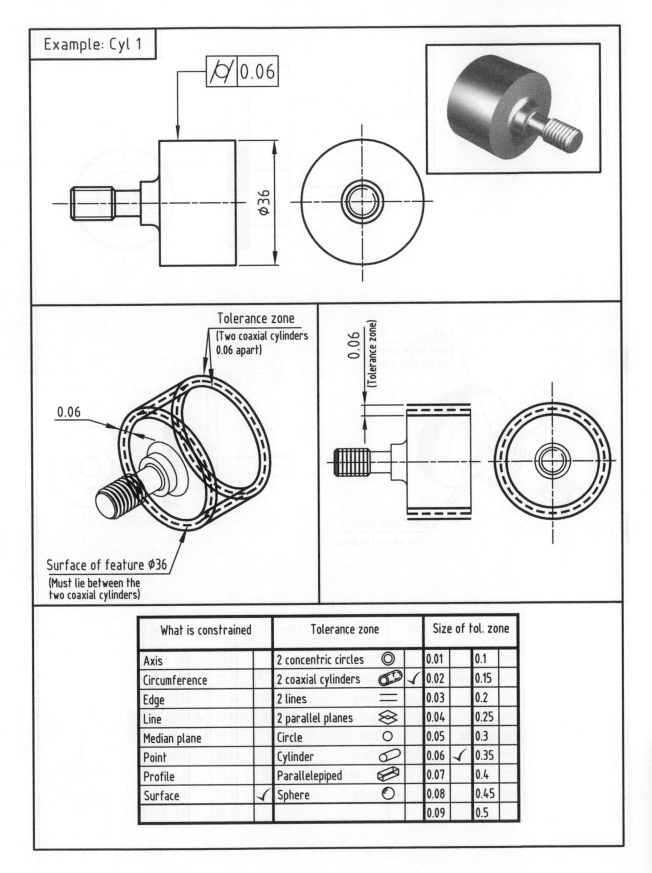

Tolerance zone
(Two coaxial cylinders
0.06 apart)

0.06

Surface of feature ⌀36
(Must lie between the
two coaxial cylinders)

0.06
(Tolerance zone)

What is constrained		Tolerance zone		Size of tol. zone		
Axis		2 concentric circles	◎	0.01		0.1
Circumference		2 coaxial cylinders	⬭ ✓	0.02		0.15
Edge		2 lines	═	0.03		0.2
Line		2 parallel planes	⬙	0.04		0.25
Median plane		Circle	○	0.05		0.3
Point		Cylinder	⬭	0.06 ✓		0.35
Profile		Parallelepiped	⬭	0.07		0.4
Surface	✓	Sphere	○	0.08		0.45
				0.09		0.5

CYLINDRICITY

Example: Fla 1

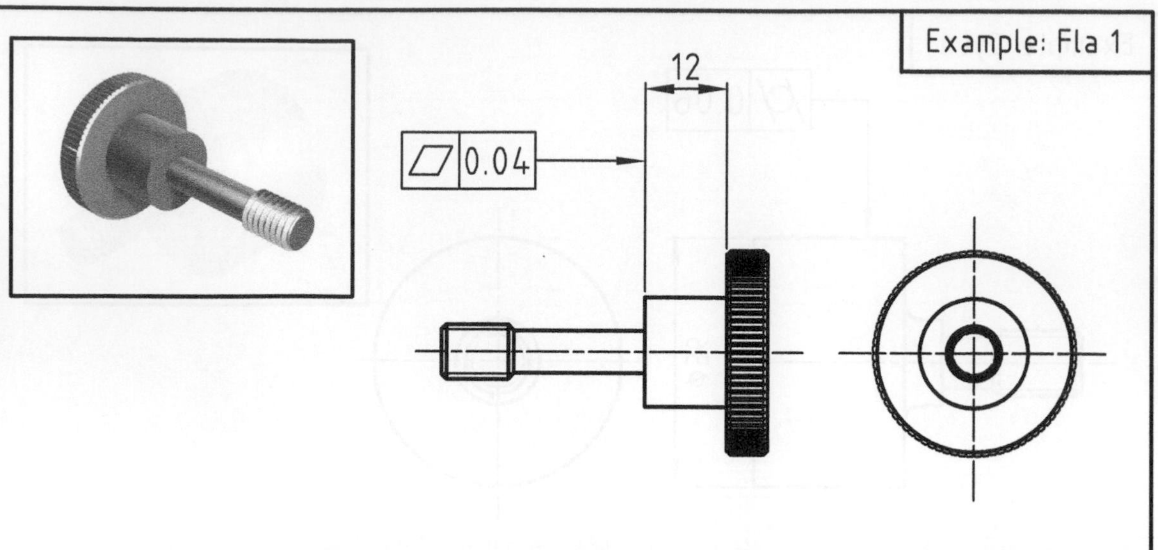
12
0.04

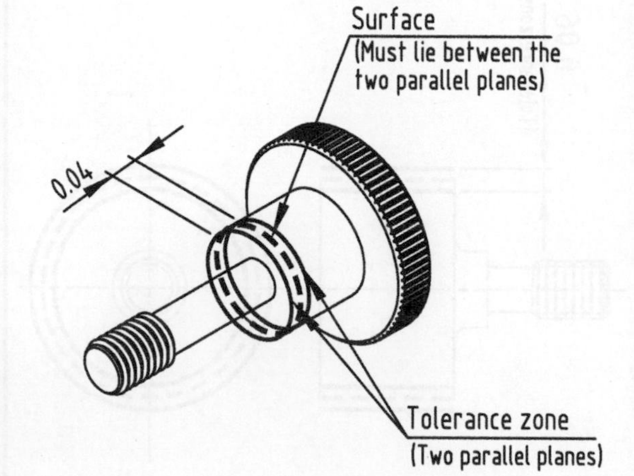
Surface
(Must lie between the
two parallel planes)
0.04
Tolerance zone
(Two parallel planes)

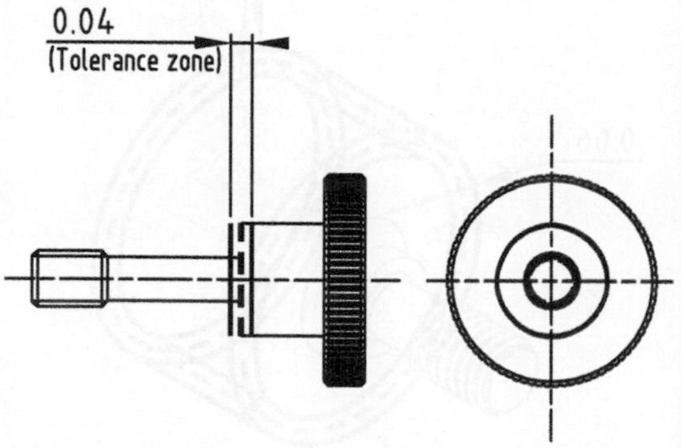
0.04
(Tolerance zone)

What is constrained		Tolerance zone		Size of tol. zone			
Axis		2 concentric circles		0.01		0.1	
Circumference		2 coaxial cylinders		0.02		0.15	
Edge		2 lines		0.03		0.2	
Line		2 parallel planes	✓	0.04	✓	0.25	
Median plane		Circle		0.05		0.3	
Point		Cylinder		0.06		0.35	
Profile		Parallelepiped		0.07		0.4	
Surface	✓	Sphere		0.08		0.45	
				0.09		0.5	

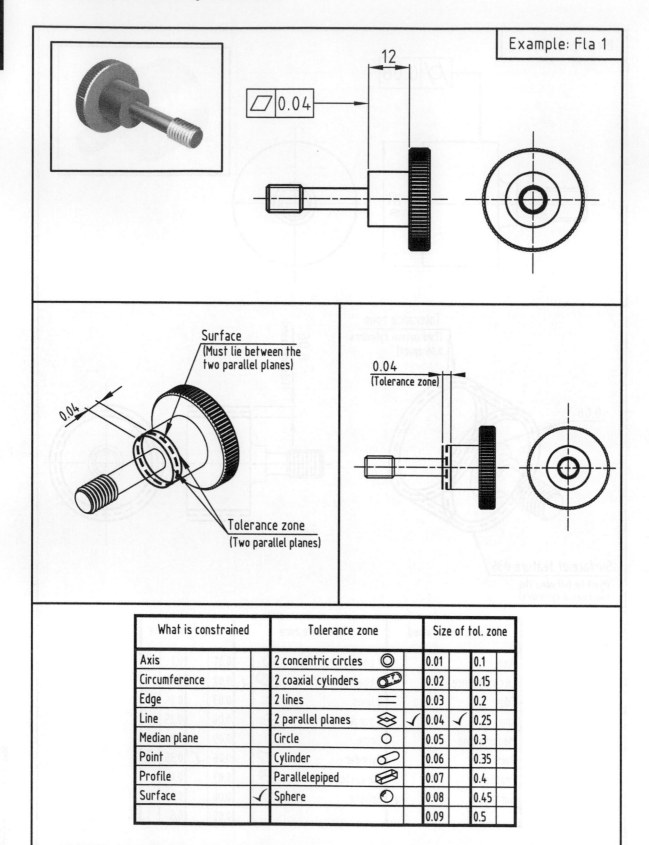

Example: Fla 2

� 0.02
NOT CONCAVE

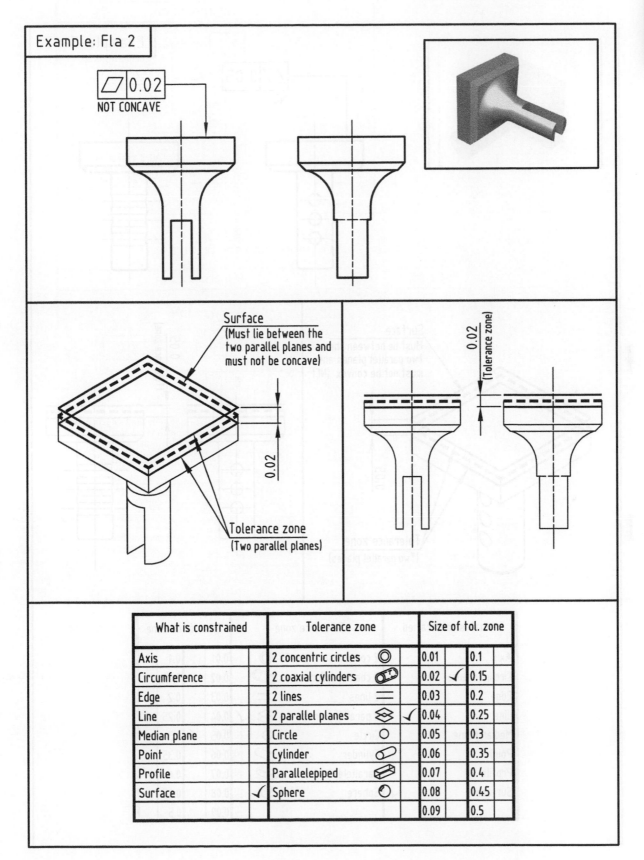

Surface
(Must lie between the
two parallel planes and
must not be concave)

0.02

Tolerance zone
(Two parallel planes)

0.02
(Tolerance zone)

What is constrained		Tolerance zone		Size of tol. zone			
Axis		2 concentric circles	◎	0.01		0.1	
Circumference		2 coaxial cylinders	⬭	0.02	✓	0.15	
Edge		2 lines	=	0.03		0.2	
Line		2 parallel planes	⬨ ✓	0.04		0.25	
Median plane		Circle	○	0.05		0.3	
Point		Cylinder	⬭	0.06		0.35	
Profile		Parallelepiped	▱	0.07		0.4	
Surface	✓	Sphere	◑	0.08		0.45	
				0.09		0.5	

Example: Fla 3

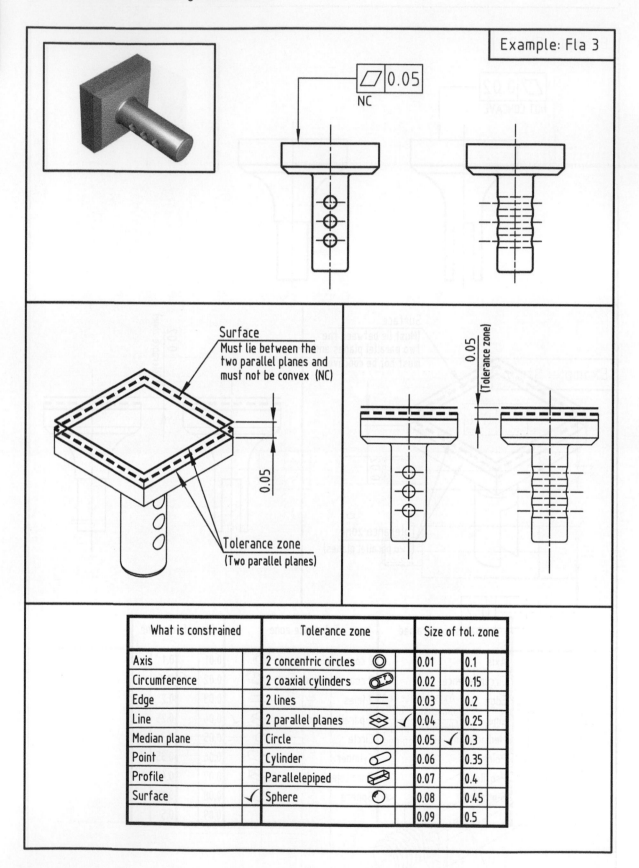

⬭ 0.05

NC

Surface
Must lie between the
two parallel planes and
must not be convex (NC)

0.05

Tolerance zone
(Two parallel planes)

0.05
(Tolerance zone)

What is constrained		Tolerance zone			Size of tol. zone		
Axis		2 concentric circles	◎		0.01		0.1
Circumference		2 coaxial cylinders			0.02		0.15
Edge		2 lines	═		0.03		0.2
Line		2 parallel planes	⬖	✓	0.04		0.25
Median plane		Circle	○		0.05	✓	0.3
Point		Cylinder	⬭		0.06		0.35
Profile		Parallelepiped	▱		0.07		0.4
Surface	✓	Sphere	◐		0.08		0.45
					0.09		0.5

Example: Par 1

// | 0.08 | A

⌀12

A

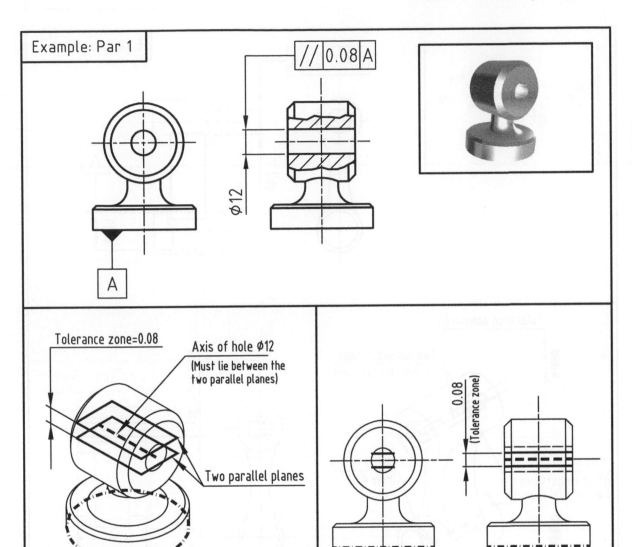

Tolerance zone=0.08

Axis of hole ⌀12
(Must lie between the two parallel planes)

Two parallel planes

Datum plane A

0.08
(Tolerance zone)

What is constrained		Tolerance zone		Size of tol. zone				Tol. zone positioned parallel to the datum	
Axis	✓	2 concentric circles	◎	0.01		0.1		Axis	
Circumference		2 coaxial cylinders	⌀	0.02		0.15		Axes	
Edge		2 lines	=	0.03		0.2		Line	
Line		2 parallel planes	⬙ ✓	0.04		0.25		Lines	
Median plane		Circle	○	0.05		0.3		Mean axis	
Point		Cylinder	⌀	0.06		0.35		Median plane	
Profile		Parallelepiped	▱	0.07		0.4		Plane A	✓
Surface		Sphere	○	0.08	✓	0.45			
				0.09		0.5			

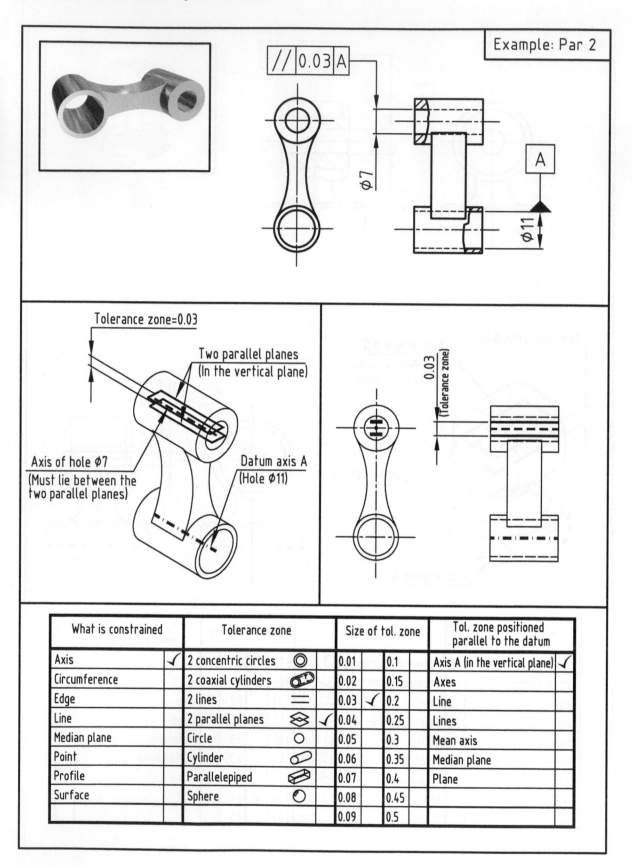

What is constrained		Tolerance zone		Size of tol. zone				Tol. zone positioned parallel to the datum	
Axis	✓	2 concentric circles	◎	0.01		0.1		Axis A (in the vertical plane)	✓
Circumference		2 coaxial cylinders	⬭	0.02		0.15		Axes	
Edge		2 lines	═	0.03	✓	0.2		Line	
Line		2 parallel planes	⬨	✓	0.04	0.25		Lines	
Median plane		Circle	○	0.05		0.3		Mean axis	
Point		Cylinder	⬭	0.06		0.35		Median plane	
Profile		Parallelepiped	▱	0.07		0.4		Plane	
Surface		Sphere	◯	0.08		0.45			
				0.09		0.5			

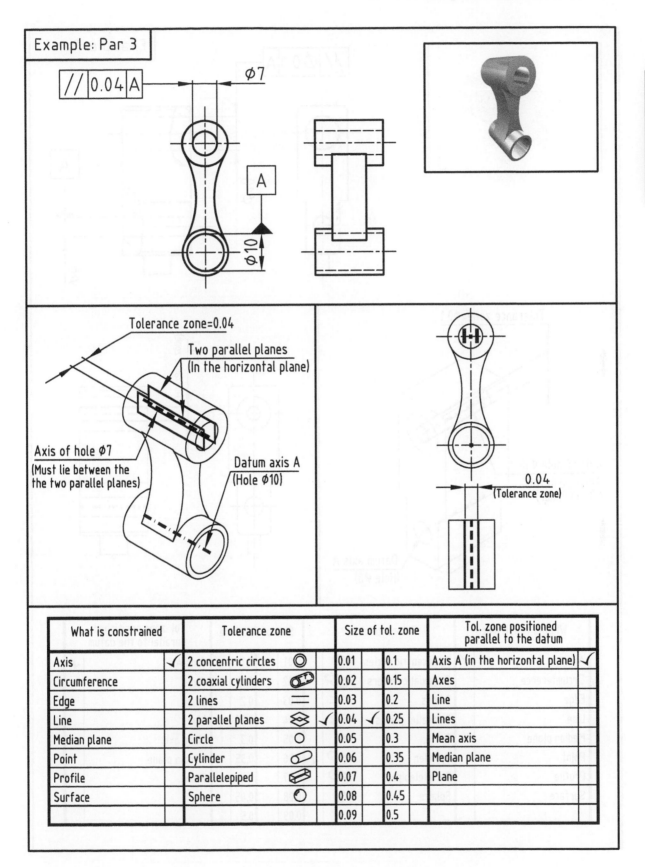

Example: Par 3

// 0.04 A

Ø7

A

Ø10

Tolerance zone=0.04

Two parallel planes
(In the horizontal plane)

Axis of hole Ø7
(Must lie between the
the two parallel planes)

Datum axis A
(Hole Ø10)

0.04
(Tolerance zone)

What is constrained		Tolerance zone		Size of tol. zone			Tol. zone positioned parallel to the datum	
Axis	✓	2 concentric circles	◎	0.01		0.1	Axis A (in the horizontal plane)	✓
Circumference		2 coaxial cylinders	⬭	0.02		0.15	Axes	
Edge		2 lines	=	0.03		0.2	Line	
Line		2 parallel planes	⬨ ✓	0.04	✓	0.25	Lines	
Median plane		Circle	○	0.05		0.3	Mean axis	
Point		Cylinder	⬭	0.06		0.35	Median plane	
Profile		Parallelepiped	⬭	0.07		0.4	Plane	
Surface		Sphere	○	0.08		0.45		
				0.09		0.5		

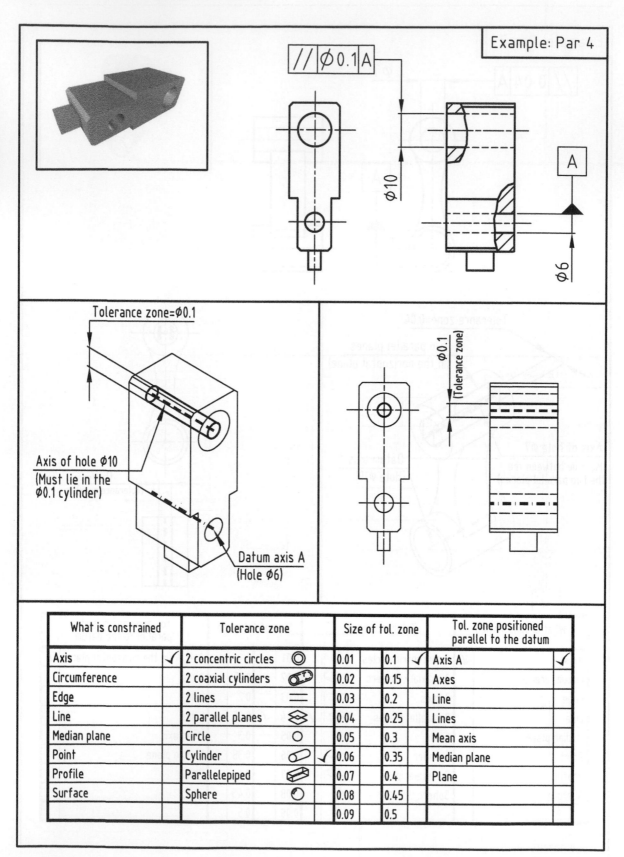

Example: Par 4

// | Ø 0.1 | A

Ø10

Ø6

A

Tolerance zone=Ø0.1

Axis of hole Ø10
(Must lie in the
Ø0.1 cylinder)

Datum axis A
(Hole Ø6)

Ø0.1
(Tolerance zone)

What is constrained		Tolerance zone		Size of tol. zone		Tol. zone positioned parallel to the datum	
Axis	✓	2 concentric circles ◎		0.01	0.1 ✓	Axis A	✓
Circumference		2 coaxial cylinders		0.02	0.15	Axes	
Edge		2 lines =		0.03	0.2	Line	
Line		2 parallel planes ⬦		0.04	0.25	Lines	
Median plane		Circle ○		0.05	0.3	Mean axis	
Point		Cylinder ⬭	✓	0.06	0.35	Median plane	
Profile		Parallelepiped ▱		0.07	0.4	Plane	
Surface		Sphere ◯		0.08	0.45		
				0.09	0.5		

PARALLELISM

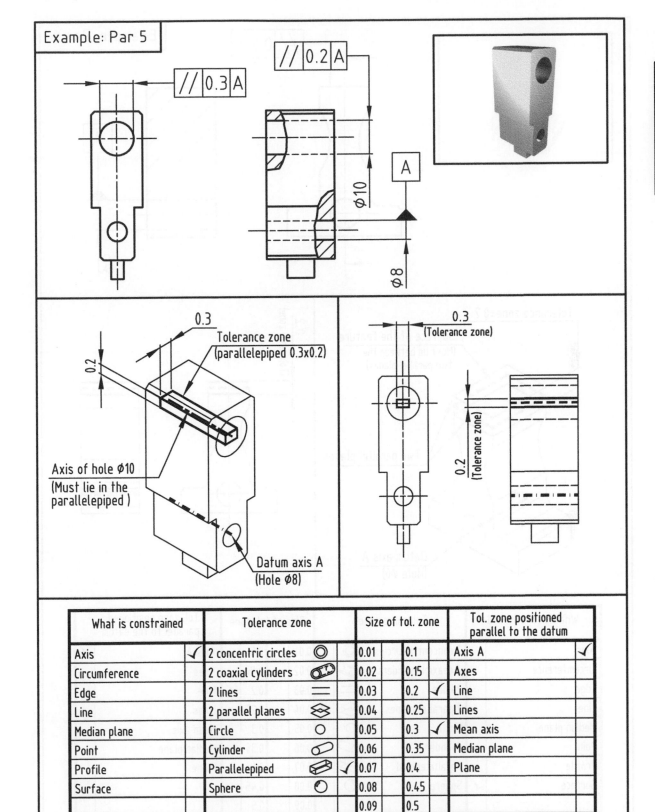

Example: Par 5

// | 0.2 | A

// | 0.3 | A

A

Ø10

Ø8

PARALLELISM

0.3

Tolerance zone
(parallelepiped 0.3x0.2)

0.2

Axis of hole Ø10
(Must lie in the
parallelepiped)

Datum axis A
(Hole Ø8)

0.3
(Tolerance zone)

0.2
(Tolerance zone)

What is constrained		Tolerance zone		Size of tol. zone		Tol. zone positioned parallel to the datum	
Axis	✓	2 concentric circles	◎	0.01	0.1	Axis A	✓
Circumference		2 coaxial cylinders	⌀	0.02	0.15	Axes	
Edge		2 lines	═	0.03	0.2 ✓	Line	
Line		2 parallel planes	⬨	0.04	0.25	Lines	
Median plane		Circle	○	0.05	0.3 ✓	Mean axis	
Point		Cylinder	⬭	0.06	0.35	Median plane	
Profile		Parallelepiped	⬨ ✓	0.07	0.4	Plane	
Surface		Sphere	⬯	0.08	0.45		
				0.09	0.5		

PARALLELISM

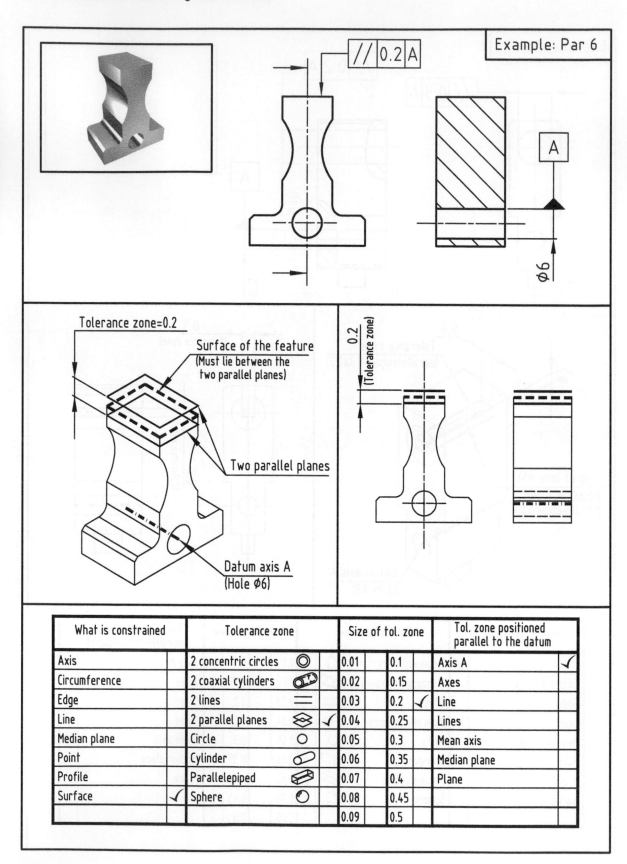

Example: Par 6

// 0.2 A

A

⌀6

Tolerance zone=0.2

Surface of the feature
(Must lie between the
two parallel planes)

Two parallel planes

Datum axis A
(Hole ⌀6)

0.2
(Tolerance zone)

What is constrained		Tolerance zone		Size of tol. zone			Tol. zone positioned parallel to the datum	
Axis		2 concentric circles	◎	0.01		0.1	Axis A	✓
Circumference		2 coaxial cylinders		0.02		0.15	Axes	
Edge		2 lines	═	0.03	✓	0.2	Line	
Line		2 parallel planes	⬦ ✓	0.04		0.25	Lines	
Median plane		Circle	○	0.05		0.3	Mean axis	
Point		Cylinder	⌁	0.06		0.35	Median plane	
Profile		Parallelepiped		0.07		0.4	Plane	
Surface	✓	Sphere	○	0.08		0.45		
				0.09		0.5		

PARALLELISM

Example: Par 7

// 0.15 A

A

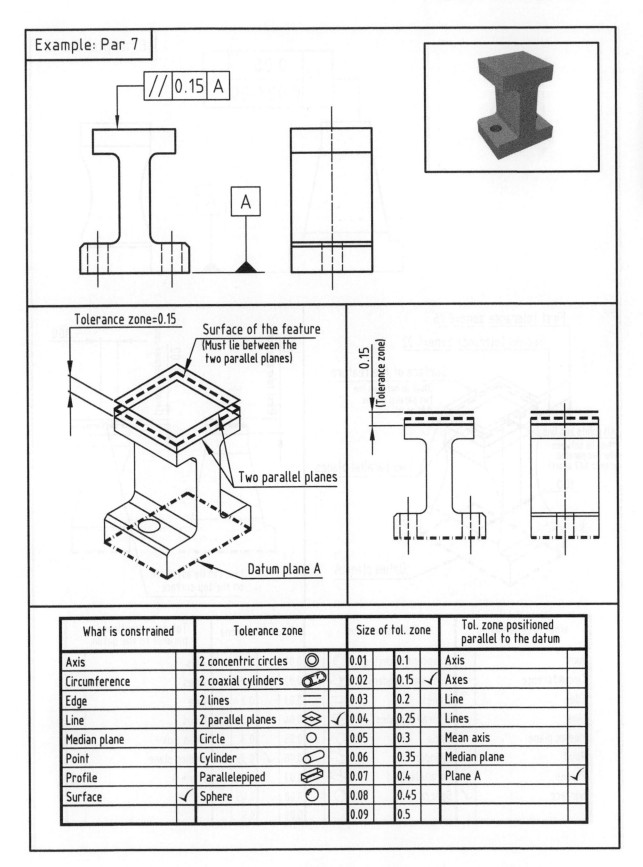

Tolerance zone=0.15

Surface of the feature
(Must lie between the
two parallel planes)

Two parallel planes

Datum plane A

0.15
(Tolerance zone)

What is constrained		Tolerance zone		Size of tol. zone			Tol. zone positioned parallel to the datum	
Axis		2 concentric circles ◎		0.01	0.1		Axis	
Circumference		2 coaxial cylinders ⬭		0.02	0.15 ✓		Axes	
Edge		2 lines =		0.03	0.2		Line	
Line		2 parallel planes ⬖ ✓		0.04	0.25		Lines	
Median plane		Circle ○		0.05	0.3		Mean axis	
Point		Cylinder ⬭		0.06	0.35		Median plane	
Profile		Parallelepiped ⬗		0.07	0.4		Plane A	✓
Surface	✓	Sphere ⬭		0.08	0.45			
				0.09	0.5			

PARALLELISM

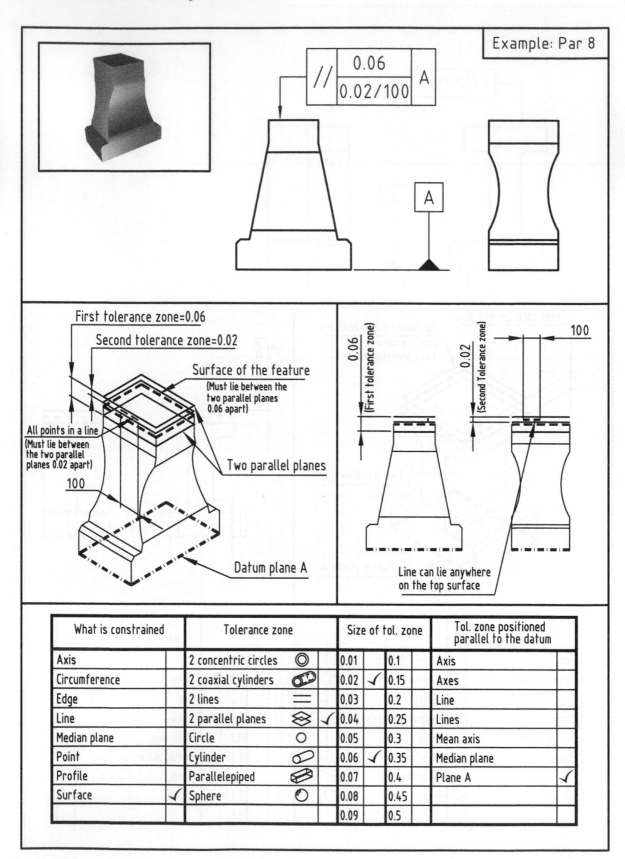

Example: Par 8

// | 0.06 / 0.02/100 | A

A

First tolerance zone=0.06

Second tolerance zone=0.02

Surface of the feature
(Must lie between the two parallel planes 0.06 apart)

All points in a line
(Must lie between the two parallel planes 0.02 apart)

Two parallel planes

100

Datum plane A

0.06 (First tolerance zone)

0.02 (Second Tolerance zone)

100

Line can lie anywhere on the top surface

What is constrained		Tolerance zone		Size of tol. zone			Tol. zone positioned parallel to the datum	
Axis		2 concentric circles	◎	0.01		0.1	Axis	
Circumference		2 coaxial cylinders	⌀	0.02	✓	0.15	Axes	
Edge		2 lines	=	0.03		0.2	Line	
Line		2 parallel planes	⬨	✓ 0.04		0.25	Lines	
Median plane		Circle	○	0.05		0.3	Mean axis	
Point		Cylinder	⌀	0.06	✓	0.35	Median plane	
Profile		Parallelepiped	▱	0.07		0.4	Plane A	✓
Surface	✓	Sphere	○	0.08		0.45		
				0.09		0.5		

PARALLELISM

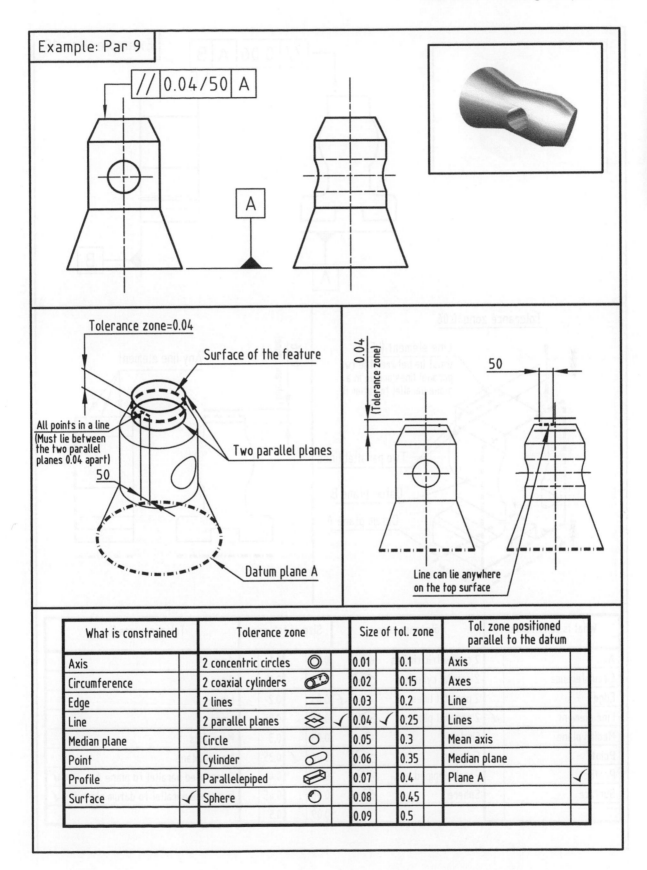

Example: Par 9

// | 0.04/50 | A

A

Tolerance zone=0.04

Surface of the feature

All points in a line
(Must lie between
the two parallel
planes 0.04 apart)

50

Two parallel planes

Datum plane A

0.04 (Tolerance zone)

50

Line can lie anywhere
on the top surface

What is constrained		Tolerance zone		Size of tol. zone			Tol. zone positioned parallel to the datum	
Axis		2 concentric circles	◎	0.01		0.1	Axis	
Circumference		2 coaxial cylinders	⌀	0.02		0.15	Axes	
Edge		2 lines	=	0.03		0.2	Line	
Line		2 parallel planes	⬨ ✓	0.04 ✓		0.25	Lines	
Median plane		Circle	○	0.05		0.3	Mean axis	
Point		Cylinder	⌀	0.06		0.35	Median plane	
Profile		Parallelepiped	▱	0.07		0.4	Plane A	✓
Surface	✓	Sphere	⬭	0.08		0.45		
				0.09		0.5		

PARALLELISM

Example: Par 10

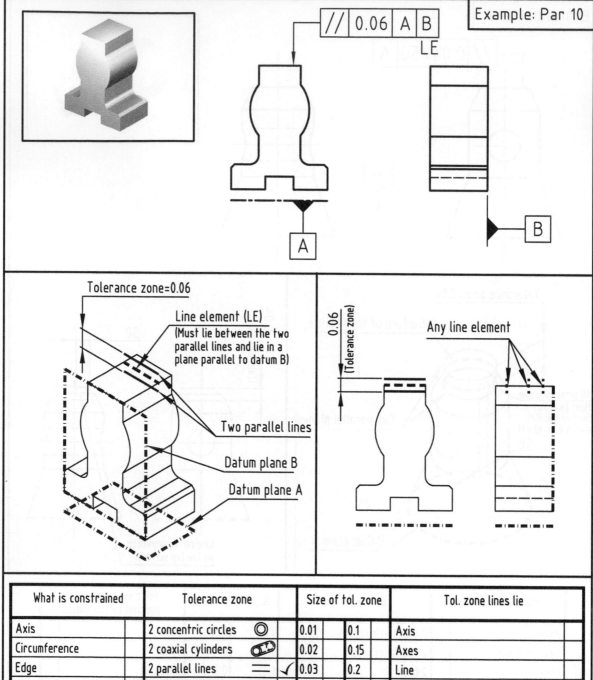

| // | 0.06 | A | B |

LE

A

B

Tolerance zone=0.06

Line element (LE)
(Must lie between the two
parallel lines and lie in a
plane parallel to datum B)

Two parallel lines

Datum plane B

Datum plane A

0.06
(Tolerance zone)

Any line element

What is constrained		Tolerance zone		Size of tol. zone		Tol. zone lines lie	
Axis		2 concentric circles ◎		0.01	0.1	Axis	
Circumference		2 coaxial cylinders		0.02	0.15	Axes	
Edge		2 parallel lines ═	✓	0.03	0.2	Line	
Line element	✓	2 parallel planes ⬦		0.04	0.25	Lines	
Median plane		Circle ○		0.05	0.3	Mean axis	
Point		Cylinder ⬭		0.06 ✓	0.35	Median plane	
Profile		Parallelepiped ▱		0.07	0.4	Orientated parallel to plane A	✓
Surface		Sphere ◖		0.08	0.45	In a plane parallel to datum plane B	✓
				0.09	0.5		

Example: Per 1

⟂ | 0.04 | A

Ø6

A

Ø7

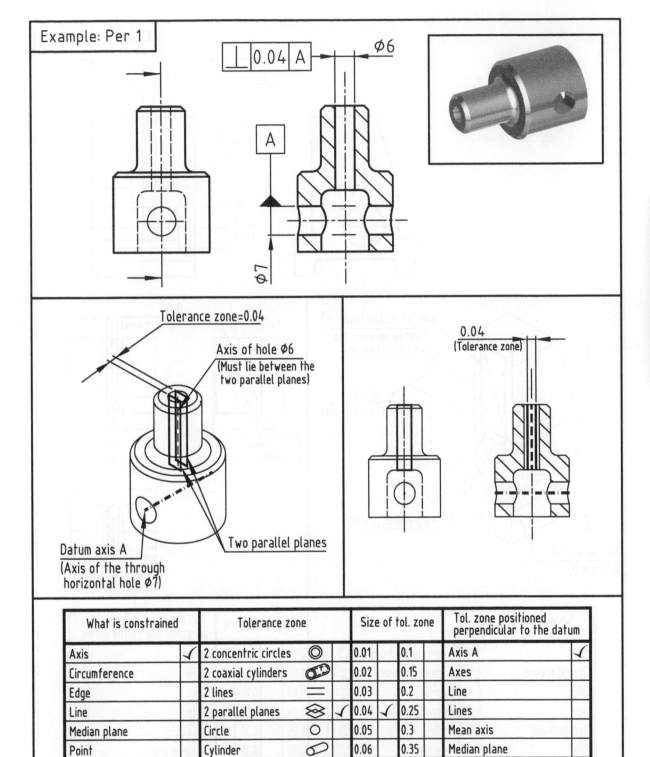

Tolerance zone=0.04

Axis of hole Ø6
(Must lie between the
two parallel planes)

Datum axis A
(Axis of the through
horizontal hole Ø7)

Two parallel planes

0.04
(Tolerance zone)

What is constrained		Tolerance zone		Size of tol. zone				Tol. zone positioned perpendicular to the datum	
Axis	✓	2 concentric circles	◎	0.01		0.1		Axis A	✓
Circumference		2 coaxial cylinders	⌀	0.02		0.15		Axes	
Edge		2 lines	=	0.03		0.2		Line	
Line		2 parallel planes	⬥	✓	0.04	✓	0.25		Lines	
Median plane		Circle	○	0.05		0.3		Mean axis	
Point		Cylinder	⌀	0.06		0.35		Median plane	
Profile		Parallelepiped	▱	0.07		0.4		Plane A	
Surface		Sphere	○	0.08		0.45			
				0.09		0.5			

PERPENDICULARITY

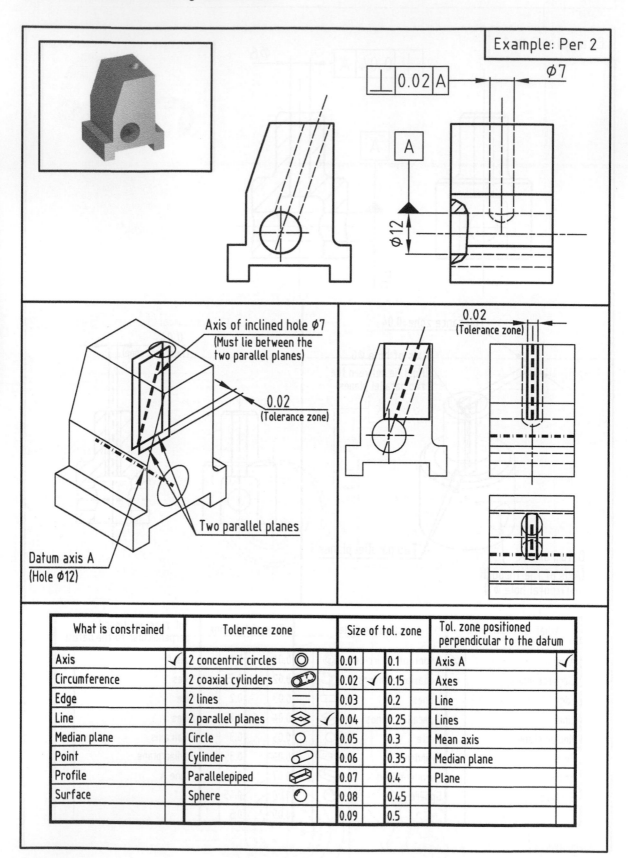

Example: Per 2

⊥ | 0.02 | A

A

∅7

∅12

Axis of inclined hole ∅7
(Must lie between the
two parallel planes)

0.02
(Tolerance zone)

Two parallel planes

Datum axis A
(Hole ∅12)

0.02
(Tolerance zone)

0.02
(Tolerance zone)

What is constrained		Tolerance zone		Size of tol. zone		Tol. zone positioned perpendicular to the datum	
Axis	✓	2 concentric circles	◎	0.01	0.1	Axis A	✓
Circumference		2 coaxial cylinders	⌀	0.02 ✓	0.15	Axes	
Edge		2 lines	═	0.03	0.2	Line	
Line		2 parallel planes	⬙ ✓	0.04	0.25	Lines	
Median plane		Circle	○	0.05	0.3	Mean axis	
Point		Cylinder	⌀	0.06	0.35	Median plane	
Profile		Parallelepiped	▱	0.07	0.4	Plane	
Surface		Sphere	⊘	0.08	0.45		
				0.09	0.5		

Example: Per 3

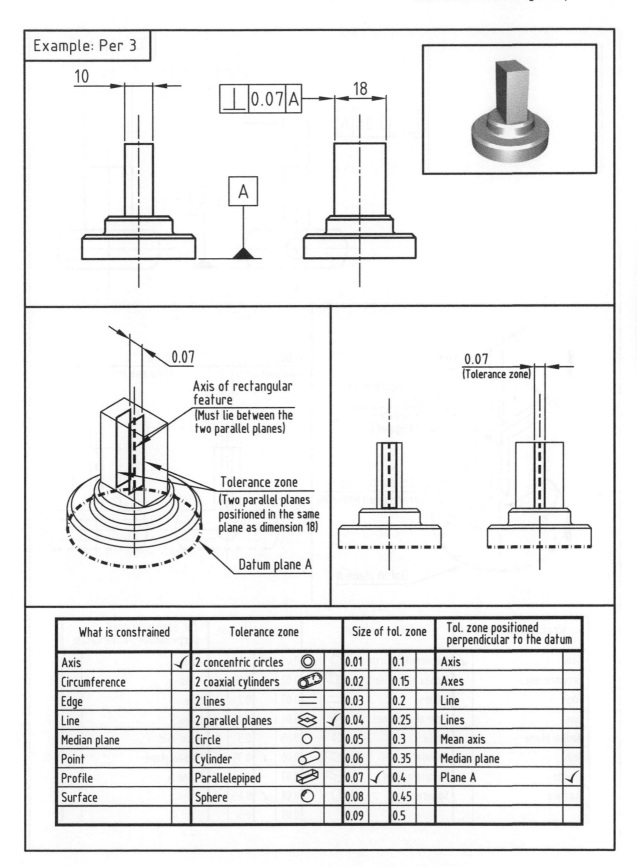

| ⊥ | 0.07 | A |

10

18

A

0.07

Axis of rectangular feature
(Must lie between the two parallel planes)

Tolerance zone
(Two parallel planes positioned in the same plane as dimension 18)

Datum plane A

0.07
(Tolerance zone)

What is constrained		Tolerance zone		Size of tol. zone				Tol. zone positioned perpendicular to the datum	
Axis	✓	2 concentric circles	◎	0.01		0.1		Axis	
Circumference		2 coaxial cylinders	⌀	0.02		0.15		Axes	
Edge		2 lines	=	0.03		0.2		Line	
Line		2 parallel planes	⬨ ✓	0.04		0.25		Lines	
Median plane		Circle	○	0.05		0.3		Mean axis	
Point		Cylinder	⌀	0.06		0.35		Median plane	
Profile		Parallelepiped	▱	0.07	✓	0.4		Plane A	✓
Surface		Sphere	○	0.08		0.45			
				0.09		0.5			

PERPENDICULARITY

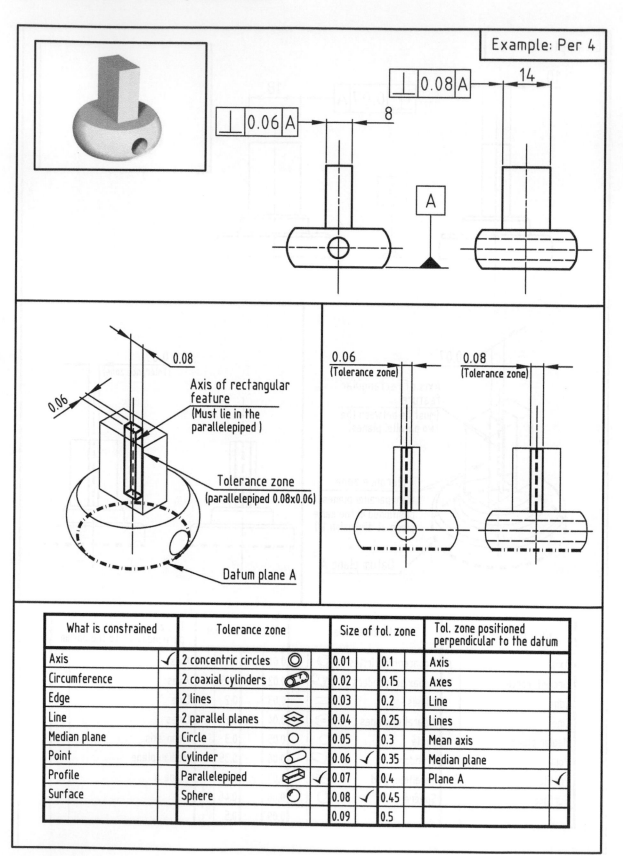

Example: Per 4

⊥ | 0.08 | A

14

⊥ | 0.06 | A

8

A

0.08

Axis of rectangular feature
(Must lie in the parallelepiped)

0.06

Tolerance zone
(parallelepiped 0.08x0.06)

Datum plane A

0.06
(Tolerance zone)

0.08
(Tolerance zone)

What is constrained		Tolerance zone		Size of tol. zone				Tol. zone positioned perpendicular to the datum	
Axis	✓	2 concentric circles	◎	0.01		0.1		Axis	
Circumference		2 coaxial cylinders		0.02		0.15		Axes	
Edge		2 lines	＝	0.03		0.2		Line	
Line		2 parallel planes	⬦	0.04		0.25		Lines	
Median plane		Circle	○	0.05		0.3		Mean axis	
Point		Cylinder	⌀	0.06	✓	0.35		Median plane	
Profile		Parallelepiped	▱	✓	0.07	0.4		Plane A	✓
Surface		Sphere	◍	0.08	✓	0.45			
				0.09		0.5			

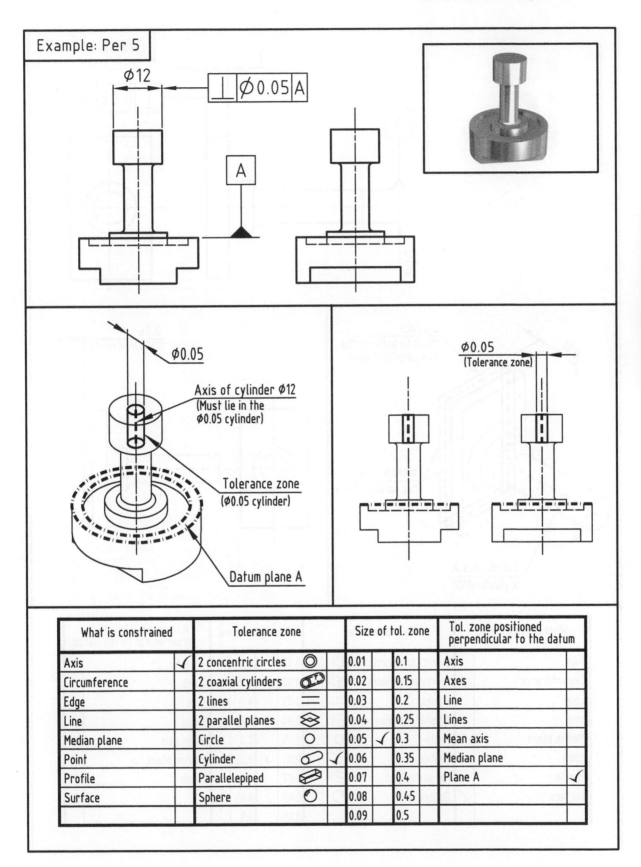

Example: Per 5

$\phi 12$

⊥ | $\phi 0.05$ | A

A

$\phi 0.05$

Axis of cylinder $\phi 12$
(Must lie in the
$\phi 0.05$ cylinder)

Tolerance zone
($\phi 0.05$ cylinder)

Datum plane A

$\phi 0.05$
(Tolerance zone)

PERPENDICULARITY

What is constrained		Tolerance zone		Size of tol. zone				Tol. zone positioned perpendicular to the datum	
Axis	✓	2 concentric circles	◎	0.01		0.1		Axis	
Circumference		2 coaxial cylinders	⌀	0.02		0.15		Axes	
Edge		2 lines	═	0.03		0.2		Line	
Line		2 parallel planes	⬨	0.04		0.25		Lines	
Median plane		Circle	○	0.05	✓	0.3		Mean axis	
Point		Cylinder	⬭	✓	0.06	0.35		Median plane	
Profile		Parallelepiped	▱	0.07		0.4		Plane A	✓
Surface		Sphere	⬯	0.08		0.45			
				0.09		0.5			

PERPENDICULARITY

Example: Per 6

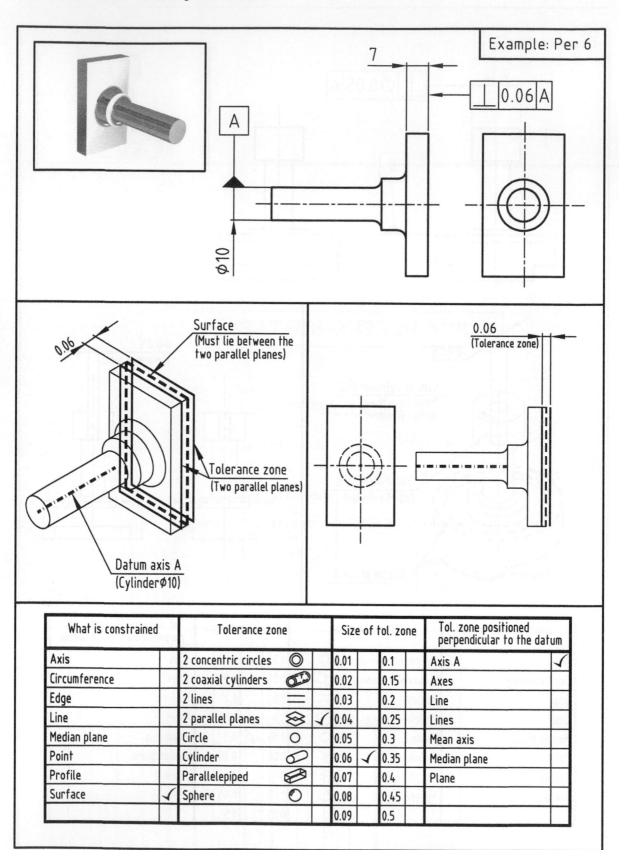

What is constrained		Tolerance zone		Size of tol. zone			Tol. zone positioned perpendicular to the datum	
Axis		2 concentric circles	◎	0.01		0.1	Axis A	✓
Circumference		2 coaxial cylinders	⌀	0.02		0.15	Axes	
Edge		2 lines	=	0.03		0.2	Line	
Line		2 parallel planes	⬗ ✓	0.04		0.25	Lines	
Median plane		Circle	○	0.05		0.3	Mean axis	
Point		Cylinder	⌀	0.06 ✓		0.35	Median plane	
Profile		Parallelepiped	▱	0.07		0.4	Plane	
Surface	✓	Sphere	⦵	0.08		0.45		
				0.09		0.5		

Example: Per 7

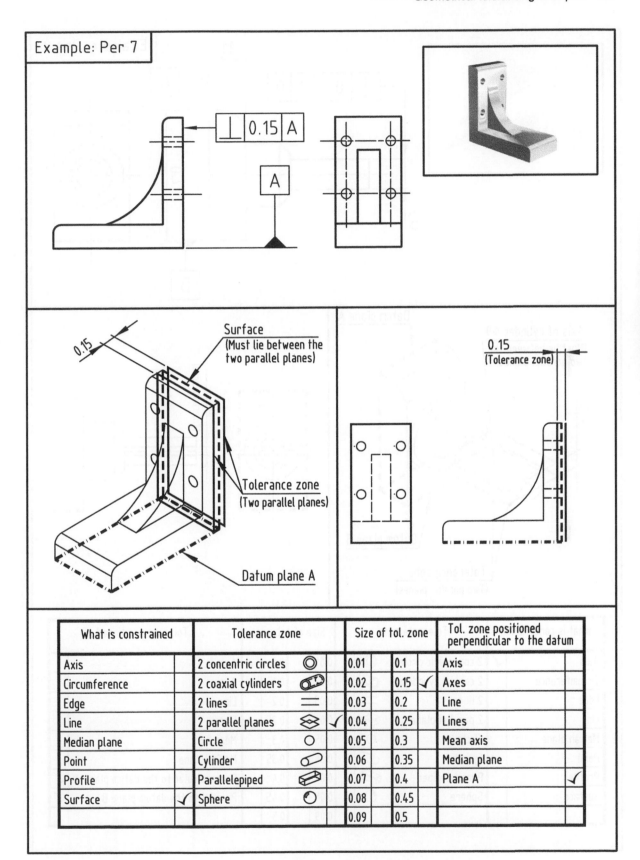

⊥ | 0.15 | A

A

0.15

Surface
(Must lie between the
two parallel planes)

Tolerance zone
(Two parallel planes)

Datum plane A

0.15
(Tolerance zone)

What is constrained		Tolerance zone		Size of tol. zone		Tol. zone positioned perpendicular to the datum	
Axis		2 concentric circles	◎	0.01	0.1	Axis	
Circumference		2 coaxial cylinders	⌀	0.02	0.15 ✓	Axes	
Edge		2 lines	=	0.03	0.2	Line	
Line		2 parallel planes ⬦	✓	0.04	0.25	Lines	
Median plane		Circle	○	0.05	0.3	Mean axis	
Point		Cylinder	⌀	0.06	0.35	Median plane	
Profile		Parallelepiped	▱	0.07	0.4	Plane A	✓
Surface	✓	Sphere	◍	0.08	0.45		
				0.09	0.5		

PERPENDICULARITY

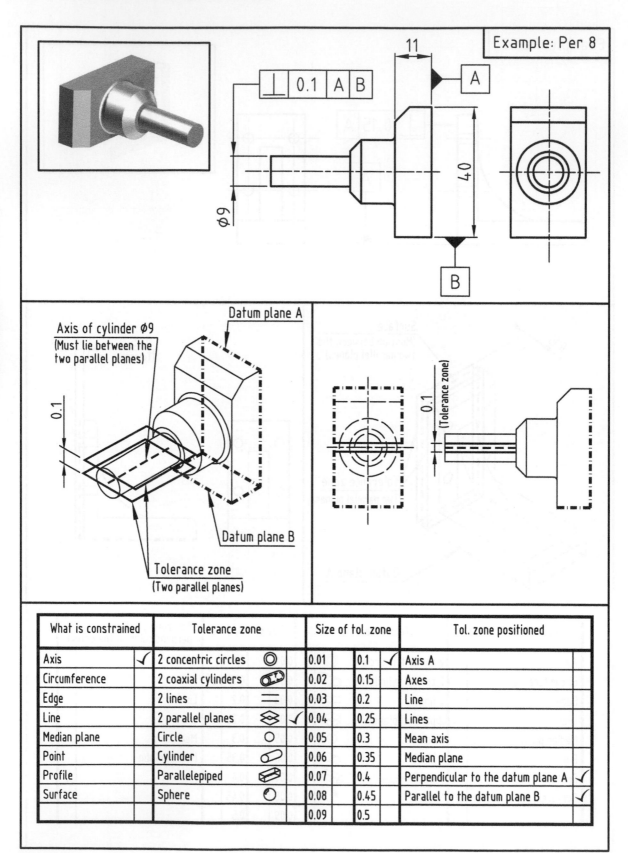

Example: Per 8

⊥ | 0.1 | A | B

A

B

Ø9

11

40

Axis of cylinder Ø9
(Must lie between the two parallel planes)

Datum plane A

0.1

Datum plane B

Tolerance zone
(Two parallel planes)

0.1
(Tolerance zone)

What is constrained		Tolerance zone		Size of tol. zone			Tol. zone positioned	
Axis	✓	2 concentric circles	◎	0.01	0.1	✓	Axis A	
Circumference		2 coaxial cylinders		0.02	0.15		Axes	
Edge		2 lines	=	0.03	0.2		Line	
Line		2 parallel planes	⬦ ✓	0.04	0.25		Lines	
Median plane		Circle	○	0.05	0.3		Mean axis	
Point		Cylinder	⌀	0.06	0.35		Median plane	
Profile		Parallelepiped	▱	0.07	0.4		Perpendicular to the datum plane A	✓
Surface		Sphere	○	0.08	0.45		Parallel to the datum plane B	✓
				0.09	0.5			

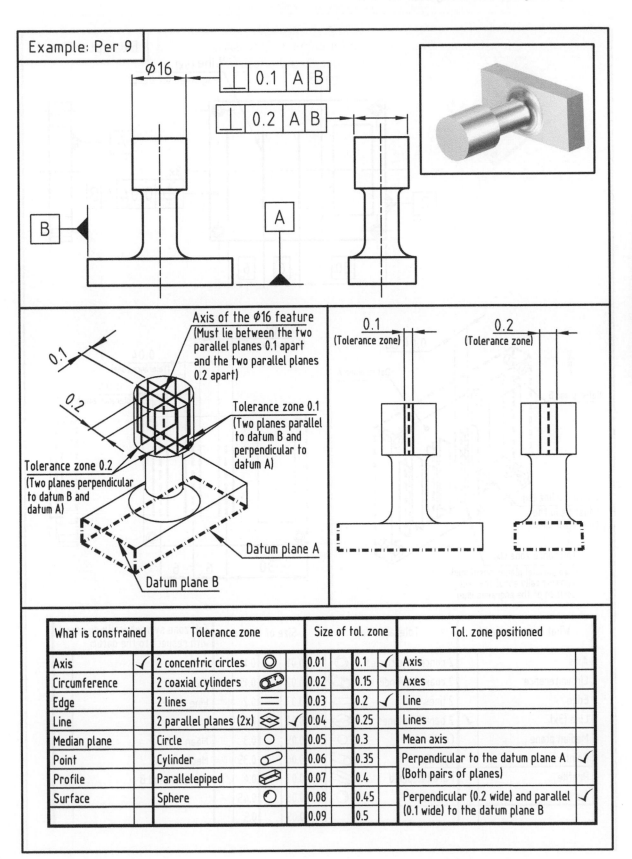

Example: Per 9

Ø16

⊥ | 0.1 | A | B

⊥ | 0.2 | A | B

B

A

Axis of the Ø16 feature
(Must lie between the two
parallel planes 0.1 apart
and the two parallel planes
0.2 apart)

0.1

0.2

Tolerance zone 0.1
(Two planes parallel
to datum B and
perpendicular to
datum A)

Tolerance zone 0.2
(Two planes perpendicular
to datum B and
datum A)

Datum plane A

Datum plane B

0.1
(Tolerance zone)

0.2
(Tolerance zone)

What is constrained		Tolerance zone		Size of tol. zone		Tol. zone positioned	
Axis	✓	2 concentric circles	◎	0.01	0.1 ✓	Axis	
Circumference		2 coaxial cylinders	⌯	0.02	0.15	Axes	
Edge		2 lines	=	0.03	0.2 ✓	Line	
Line		2 parallel planes (2x) ✓	⬦	0.04	0.25	Lines	
Median plane		Circle	○	0.05	0.3	Mean axis	
Point		Cylinder	⌀	0.06	0.35	Perpendicular to the datum plane A	✓
Profile		Parallelepiped	▱	0.07	0.4	(Both pairs of planes)	
Surface		Sphere	○	0.08	0.45	Perpendicular (0.2 wide) and parallel	✓
				0.09	0.5	(0.1 wide) to the datum plane B	

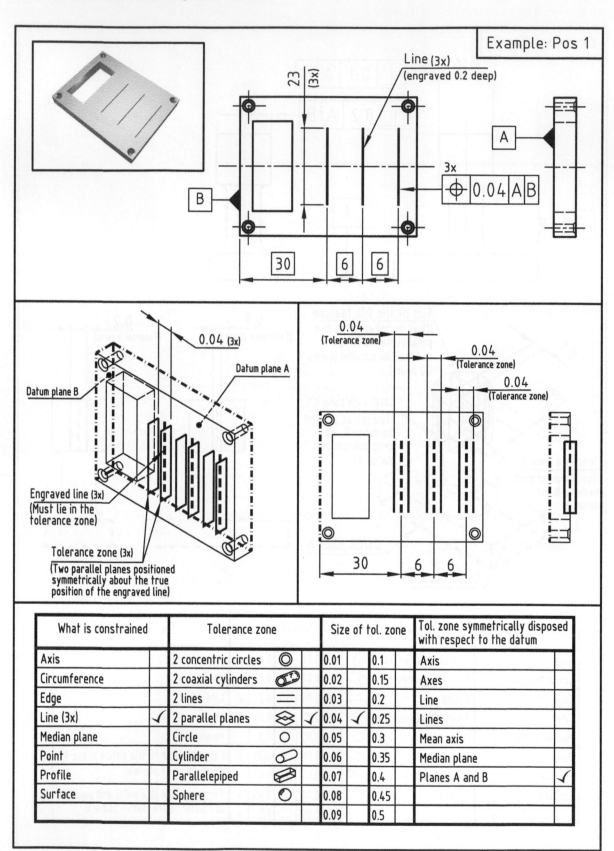

Example: Pos 1

23 (3x)

Line (3x)
(engraved 0.2 deep)

A

3x

⊕ | 0.04 | A | B

B

30   6   6

0.04 (3x)

Datum plane A

Datum plane B

Engraved line (3x)
(Must lie in the
tolerance zone)

Tolerance zone (3x)
(Two parallel planes positioned
symmetrically about the true
position of the engraved line)

0.04
(Tolerance zone)

0.04
(Tolerance zone)

0.04
(Tolerance zone)

30   6   6

What is constrained	Tolerance zone		Size of tol. zone			Tol. zone symmetrically disposed with respect to the datum	
Axis	2 concentric circles	◎	0.01		0.1	Axis	
Circumference	2 coaxial cylinders	⌭	0.02		0.15	Axes	
Edge	2 lines	═	0.03		0.2	Line	
Line (3x) ✓	2 parallel planes	⬙ ✓	0.04 ✓		0.25	Lines	
Median plane	Circle	○	0.05		0.3	Mean axis	
Point	Cylinder	⌀	0.06		0.35	Median plane	
Profile	Parallelepiped	▱	0.07		0.4	Planes A and B	✓
Surface	Sphere	◑	0.08		0.45		
			0.09		0.5		

POSITION

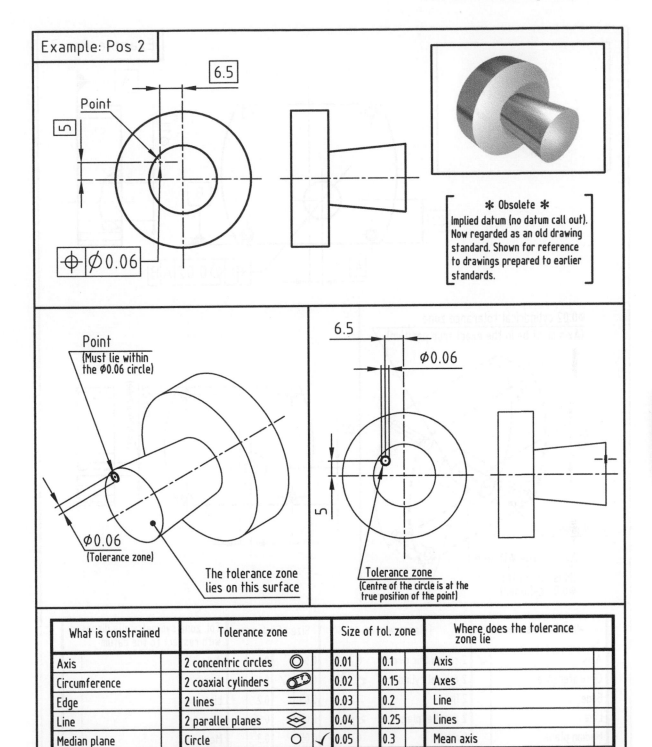

Example: Pos 2

6.5

5

Point

⊕ ⌀0.06

∗ Obsolete ∗
Implied datum (no datum call out).
Now regarded as an old drawing
standard. Shown for reference
to drawings prepared to earlier
standards.

Point
(Must lie within
the ⌀0.06 circle)

⌀0.06
(Tolerance zone)

The tolerance zone
lies on this surface

6.5

⌀0.06

5

Tolerance zone
(Centre of the circle is at the
true position of the point)

POSITION

What is constrained		Tolerance zone		Size of tol. zone		Where does the tolerance zone lie	
Axis		2 concentric circles ◎		0.01	0.1	Axis	
Circumference		2 coaxial cylinders		0.02	0.15	Axes	
Edge		2 lines ═		0.03	0.2	Line	
Line		2 parallel planes ◈		0.04	0.25	Lines	
Median plane		Circle ○	✓	0.05	0.3	Mean axis	
Point	✓	Cylinder		0.06 ✓	0.35	Median plane	
Profile		Parallelepiped		0.07	0.4	Plane	
Surface		Sphere		0.08	0.45		
				0.09	0.5	Centred at the true position of the point	✓

Example: Pos 3

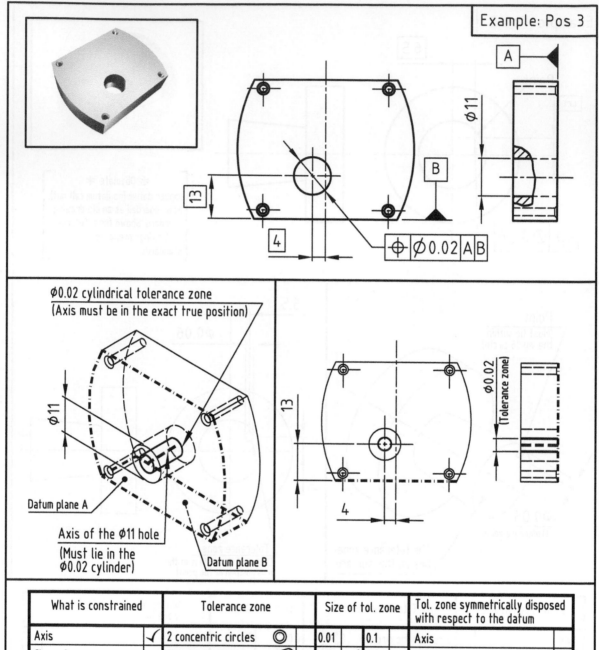

φ0.02 cylindrical tolerance zone
(Axis must be in the exact true position)

Datum plane A

Axis of the φ11 hole
(Must lie in the
φ0.02 cylinder)

Datum plane B

What is constrained	Tolerance zone		Size of tol. zone		Tol. zone symmetrically disposed with respect to the datum	
Axis	✓ 2 concentric circles	◎	0.01	0.1	Axis	
Circumference	2 coaxial cylinders	⬭ ✓	0.02	0.15	Axes	
Edge	2 lines	=	0.03	0.2	Line	
Line	2 parallel planes	⬥	0.04	0.25	Lines	
Median plane	Circle	○	0.05	0.3	Mean axis	
Point	Cylinder	⬮ ✓	0.06	0.35	Median plane	
Profile	Parallelepiped	▱	0.07	0.4	Planes A and B	✓
Surface	Sphere	◓	0.08	0.45		
			0.09	0.5		

POSITION

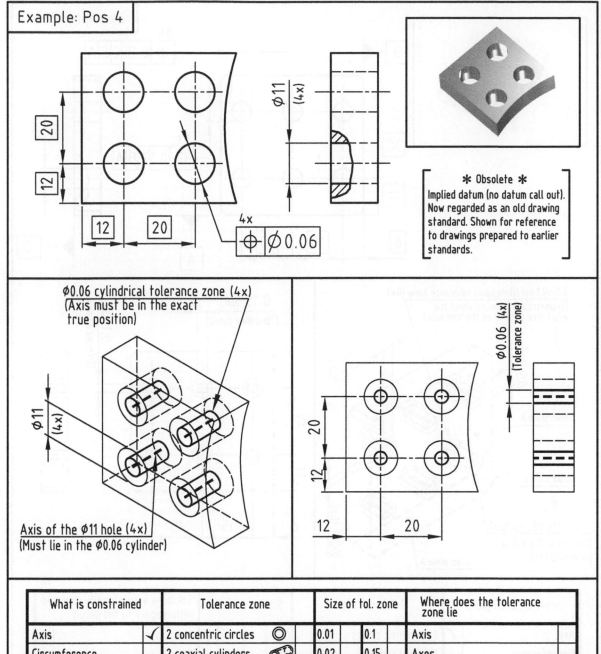

Example: Pos 4

Ø11 (4x)

4x
⊕ Ø0.06

* Obsolete *
Implied datum (no datum call out).
Now regarded as an old drawing
standard. Shown for reference
to drawings prepared to earlier
standards.

Ø0.06 cylindrical tolerance zone (4x)
(Axis must be in the exact
true position)

Ø11 (4x)

Axis of the Ø11 hole (4x)
(Must lie in the Ø0.06 cylinder)

Ø0.06 (4x)
(Tolerance zone)

What is constrained		Tolerance zone		Size of tol. zone		Where does the tolerance zone lie	
Axis	✓	2 concentric circles	◎	0.01		Axis	
Circumference		2 coaxial cylinders		0.02		Axes	
Edge		2 lines	=	0.03		Line	
Line		2 parallel planes		0.04		Lines	
Median plane		Circle	○	0.05		Mean axis	
Point		Cylinder	▱ ✓	0.06 ✓		Median plane	
Profile		Parallelepiped		0.07		Plane	
Surface		Sphere	◔	0.08		Its axis is to be coincident with	
				0.09	0.5	true position of the Ø11 hole axis	✓

Size of tol. zone second column values: 0.1, 0.15, 0.2, 0.25, 0.3, 0.35, 0.4, 0.45, 0.5

POSITION

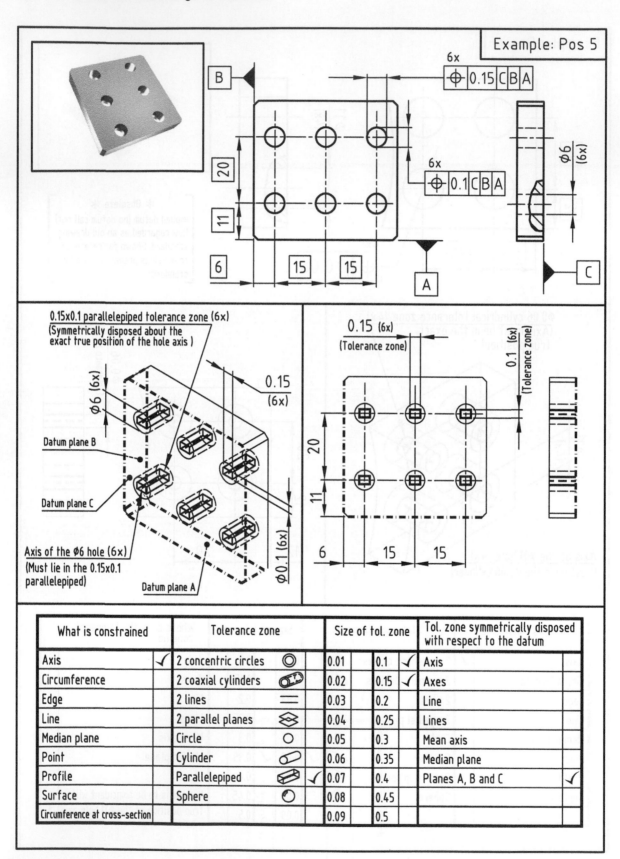

Example: Pos 5

What is constrained		Tolerance zone		Size of tol. zone			Tol. zone symmetrically disposed with respect to the datum	
Axis	✓	2 concentric circles	◎	0.01	0.1	✓	Axis	
Circumference		2 coaxial cylinders		0.02	0.15	✓	Axes	
Edge		2 lines	═	0.03	0.2		Line	
Line		2 parallel planes	⬙	0.04	0.25		Lines	
Median plane		Circle	○	0.05	0.3		Mean axis	
Point		Cylinder	⌀	0.06	0.35		Median plane	
Profile		Parallelepiped	▱	✓	0.07	0.4	Planes A, B and C	✓
Surface		Sphere	◑	0.08	0.45			
Circumference at cross-section				0.09	0.5			

Example: Pos 6

⊕ S⌀0.1 A B C

Red plastic insert

Clear plastic

12

19

B

16

C

Point

A

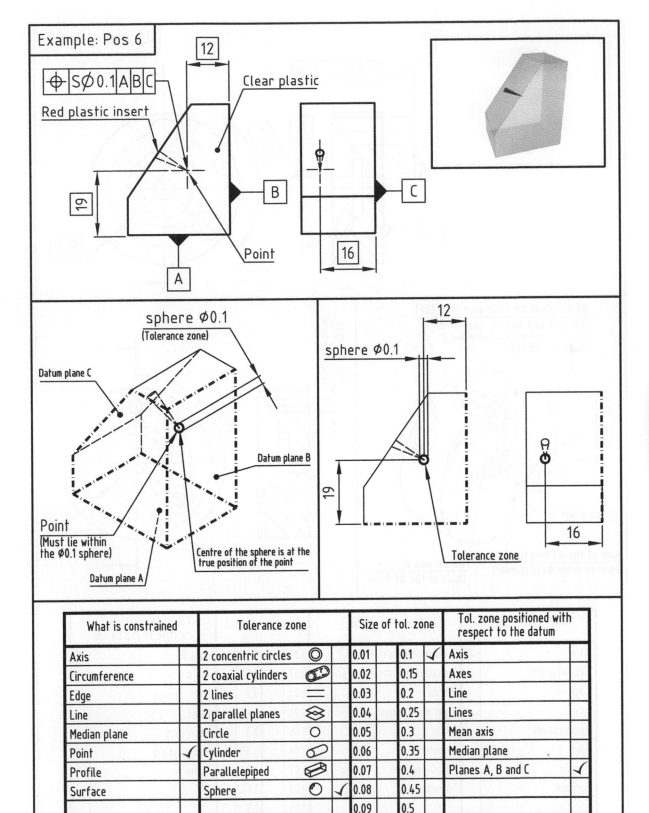

sphere ⌀0.1
(Tolerance zone)

Datum plane C

Datum plane B

Point
(Must lie within
the ⌀0.1 sphere)

Centre of the sphere is at the
true position of the point

Datum plane A

sphere ⌀0.1

12

19

16

Tolerance zone

POSITION

What is constrained		Tolerance zone		Size of tol. zone		Tol. zone positioned with respect to the datum	
Axis		2 concentric circles ◎		0.01	0.1 ✓	Axis	
Circumference		2 coaxial cylinders		0.02	0.15	Axes	
Edge		2 lines ═		0.03	0.2	Line	
Line		2 parallel planes ⬨		0.04	0.25	Lines	
Median plane		Circle ○		0.05	0.3	Mean axis	
Point	✓	Cylinder ⬭		0.06	0.35	Median plane	
Profile		Parallelepiped ⬓		0.07	0.4	Planes A, B and C	✓
Surface		Sphere ◯	✓	0.08	0.45		
				0.09	0.5		

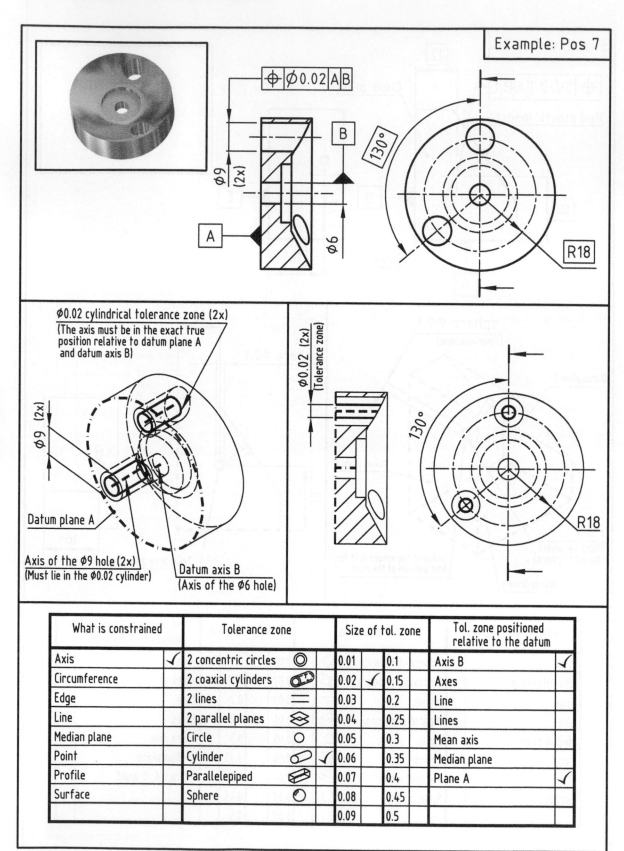

Example: Pos 7

⊕ | ⌀0.02 | A | B

⌀9 (2x)

B

130°

A

⌀6

R18

⌀0.02 cylindrical tolerance zone (2x)
(The axis must be in the exact true
position relative to datum plane A
and datum axis B)

⌀9 (2x)

Datum plane A

Axis of the ⌀9 hole (2x)
(Must lie in the ⌀0.02 cylinder)

Datum axis B
(Axis of the ⌀6 hole)

⌀0.02 (2x)
(Tolerance zone)

130°

R18

What is constrained		Tolerance zone		Size of tol. zone				Tol. zone positioned relative to the datum	
Axis	✓	2 concentric circles	◎	0.01		0.1		Axis B	✓
Circumference		2 coaxial cylinders	⌀	0.02	✓	0.15		Axes	
Edge		2 lines	═	0.03		0.2		Line	
Line		2 parallel planes	◈	0.04		0.25		Lines	
Median plane		Circle	○	0.05		0.3		Mean axis	
Point		Cylinder	⌀	✓ 0.06		0.35		Median plane	
Profile		Parallelepiped	▱	0.07		0.4		Plane A	✓
Surface		Sphere	◯	0.08		0.45			
				0.09		0.5			

POSITION

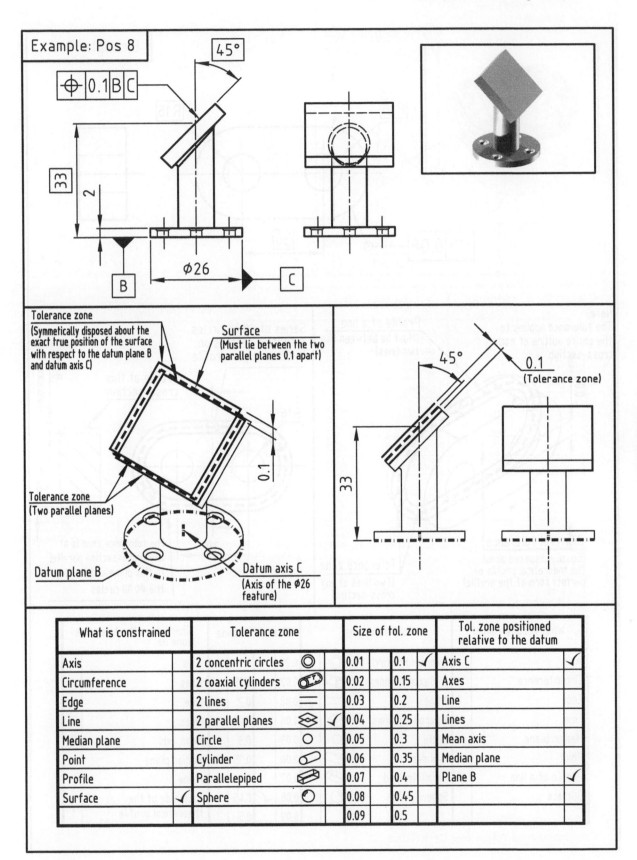

What is constrained		Tolerance zone		Size of tol. zone			Tol. zone positioned relative to the datum	
Axis		2 concentric circles	◎	0.01	0.1	✓	Axis C	✓
Circumference		2 coaxial cylinders	⌀	0.02	0.15		Axes	
Edge		2 lines	═	0.03	0.2		Line	
Line		2 parallel planes	⬦	✓0.04	0.25		Lines	
Median plane		Circle	○	0.05	0.3		Mean axis	
Point		Cylinder	⬭	0.06	0.35		Median plane	
Profile		Parallelepiped	▱	0.07	0.4		Plane B	✓
Surface	✓	Sphere	◍	0.08	0.45			
				0.09	0.5			

POSITION

Example: Pos 8

45°

⊕ 0.1 B C

33

2

Ø26

B

C

Tolerance zone (Symmetrically disposed about the exact true position of the surface with respect to the datum plane B and datum axis C)

Surface (Must lie between the two parallel planes 0.1 apart)

Tolerance zone (Two parallel planes)

Datum plane B

Datum axis C (Axis of the Ø26 feature)

45°

0.1 (Tolerance zone)

33

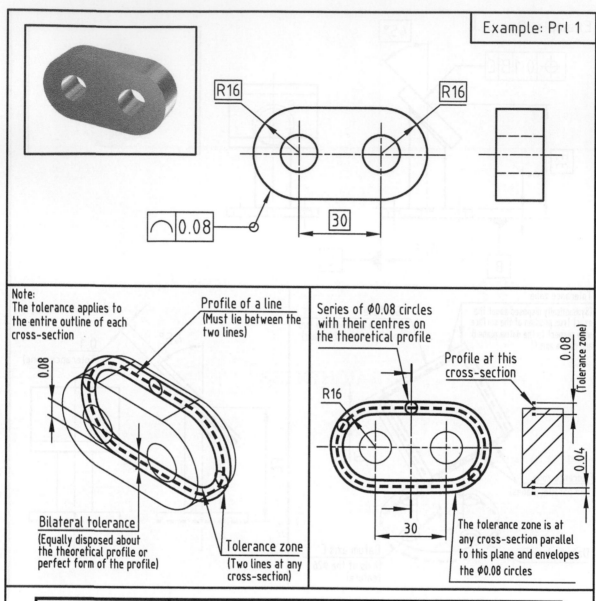

Example: Prl 1

R16   R16

⌒ 0.08

30

Note:
The tolerance applies to the entire outline of each cross-section

0.08

Profile of a line
(Must lie between the two lines)

Bilateral tolerance
(Equally disposed about the theoretical profile or perfect form of the profile)

Tolerance zone
(Two lines at any cross-section)

Series of Ø0.08 circles with their centres on the theoretical profile

R16

Profile at this cross-section

0.08 (Tolerance zone)

0.04

30

The tolerance zone is at any cross-section parallel to this plane and envelopes the Ø0.08 circles

What is constrained		Tolerance zone		Size of tol. zone		Where does the tolerance zone lie	
Axis		2 concentric circles ◎		0.01	0.1	Axis	
Circumference		2 coaxial cylinders		0.02	0.15	Axes	
Edge		2 lines ═	✓	0.03	0.2	Line	
Line		2 parallel planes ⬙		0.04	0.25	Lines	
Median plane		Circle ○		0.05	0.3	Mean axis	
Point		Cylinder ⬭		0.06	0.35	Median plane	
Profile of a line	✓	Parallelepiped ▱		0.07	0.4	Plane	
Surface		Sphere ⬤		0.08 ✓	0.45	Each side of the theoretical profile	✓
				0.09	0.5		

PROFILE OF A LINE

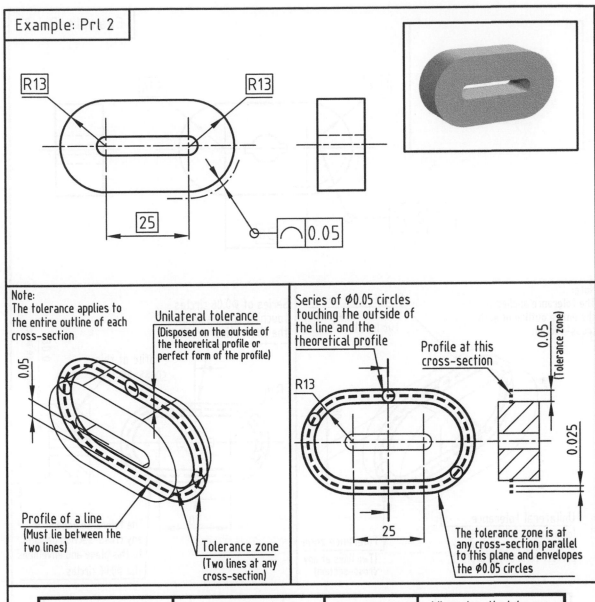

Example: Prl 2

R13   R13

25

⌒ 0.05

Note:
The tolerance applies to the entire outline of each cross-section

Unilateral tolerance
(Disposed on the outside of the theoretical profile or perfect form of the profile)

0.05

Profile of a line
(Must lie between the two lines)

Tolerance zone
(Two lines at any cross-section)

Series of Ø0.05 circles touching the outside of the line and the theoretical profile

R13

Profile at this cross-section

0.05 (Tolerance zone)

0.025

25

The tolerance zone is at any cross-section parallel to this plane and envelopes the Ø0.05 circles

What is constrained		Tolerance zone		Size of tol. zone		Where does the tolerance zone lie	
Axis		2 concentric circles	◎	0.01	0.1	Axis	
Circumference		2 coaxial cylinders		0.02	0.15	Axes	
Edge		2 lines	= ✓	0.03	0.2	Line	
Line		2 parallel planes	⬨	0.04	0.25	Lines	
Median plane		Circle	○	0.05 ✓	0.3	Mean axis	
Point		Cylinder	⌀	0.06	0.35	Median plane	
Profile of a line	✓	Parallelepiped	▱	0.07	0.4	Plane	
Surface		Sphere	◉	0.08	0.45	Outside of the theoretical profile	✓
				0.09	0.5		

PROFILE OF A LINE

Example: Prl 3

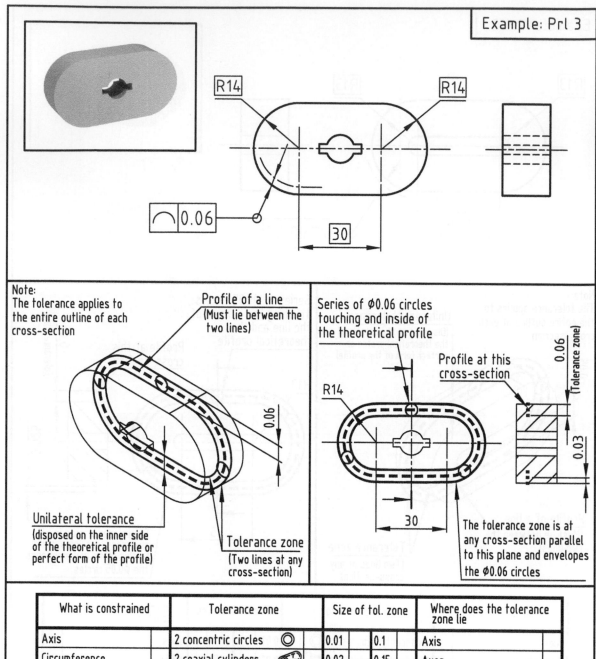

Note:
The tolerance applies to the entire outline of each cross-section

Profile of a line
(Must lie between the two lines)

Unilateral tolerance
(disposed on the inner side of the theoretical profile or perfect form of the profile)

Tolerance zone
(Two lines at any cross-section)

Series of Ø0.06 circles touching and inside of the theoretical profile

Profile at this cross-section

The tolerance zone is at any cross-section parallel to this plane and envelopes the Ø0.06 circles

What is constrained		Tolerance zone		Size of tol. zone		Where does the tolerance zone lie	
Axis		2 concentric circles ◎		0.01	0.1	Axis	
Circumference		2 coaxial cylinders ⬭		0.02	0.15	Axes	
Edge		2 lines ═	✓	0.03	0.2	Line	
Line		2 parallel planes ⬥		0.04	0.25	Lines	
Median plane		Circle ○		0.05	0.3	Mean axis	
Point		Cylinder ⬭		0.06 ✓	0.35	Median plane	
Profile of a line	✓	Parallelepiped ⬦		0.07	0.4	Plane	
Surface		Sphere ◉		0.08	0.45	Inner side of the theoretical profile	✓
				0.09	0.5		

PROFILE OF A LINE

Example: Prl 4

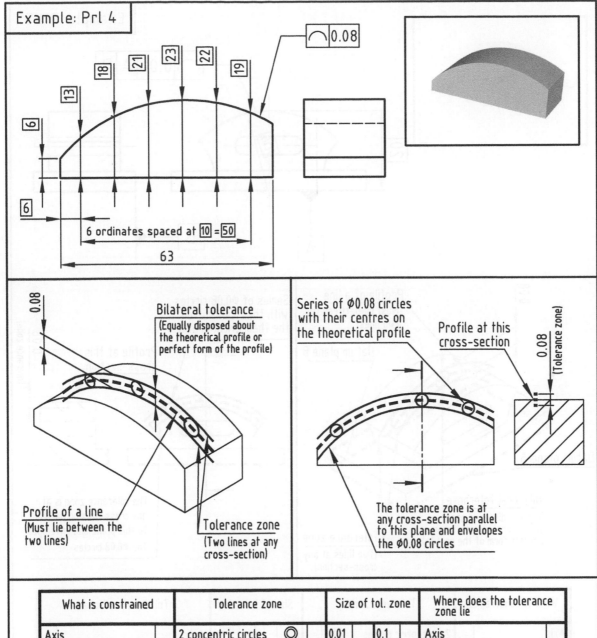

⌒ 0.08

6 ordinates spaced at [10] = [50]

63

0.08

Bilateral tolerance
(Equally disposed about
the theoretical profile or
perfect form of the profile)

Series of ⌀0.08 circles
with their centres on
the theoretical profile

Profile at this
cross-section

0.08
(Tolerance zone)

Profile of a line
(Must lie between the
two lines)

Tolerance zone
(Two lines at any
cross-section)

The tolerance zone is at
any cross-section parallel
to this plane and envelopes
the ⌀0.08 circles

What is constrained		Tolerance zone		Size of tol. zone			Where does the tolerance zone lie	
Axis		2 concentric circles ◎		0.01		0.1	Axis	
Circumference		2 coaxial cylinders		0.02		0.15	Axes	
Edge		2 lines =	✓	0.03		0.2	Line	
Line		2 parallel planes ⬦		0.04		0.25	Lines	
Median plane		Circle ○		0.05		0.3	Mean axis	
Point		Cylinder ⬭		0.06		0.35	Median plane	
Profile of a line	✓	Parallelepiped ⬙		0.07		0.4	Plane	
Surface		Sphere ⬭		0.08	✓	0.45	Each side of the theoretical profile	✓
				0.09		0.5		

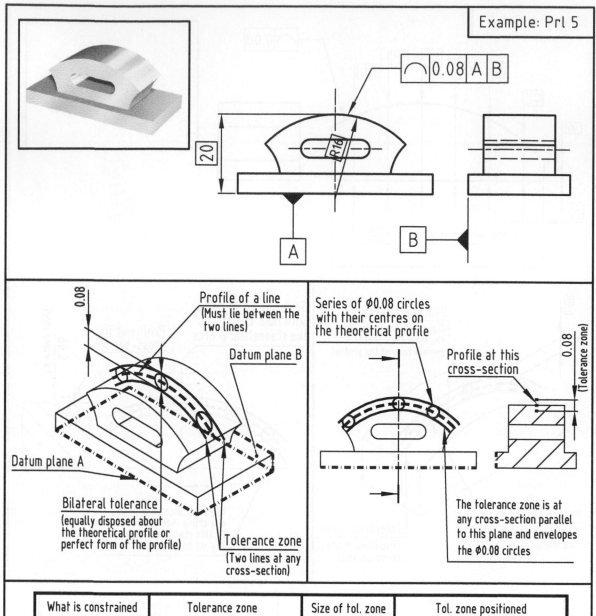

Example: Prl 5

$\frown$ 0.08 A B

20

R16

A

B

Profile of a line
(Must lie between the
two lines)

Datum plane B

0.08

Datum plane A

Bilateral tolerance
(equally disposed about
the theoretical profile or
perfect form of the profile)

Tolerance zone
(Two lines at any
cross-section)

Series of $\emptyset$0.08 circles
with their centres on
the theoretical profile

Profile at this
cross-section

0.08
(Tolerance zone)

The tolerance zone is at
any cross-section parallel
to this plane and envelopes
the $\emptyset$0.08 circles

PROFILE OF A LINE

What is constrained		Tolerance zone		Size of tol. zone				Tol. zone positioned	
Axis		2 concentric circles	◎	0.01		0.1		Axis	
Circumference		2 coaxial cylinders		0.02		0.15		Axes	
Edge		2 lines	= ✓	0.03		0.2		Line	
Line		2 parallel planes	⬦	0.04		0.25		Lines	
Median plane		Circle	○	0.05		0.3		Mean axis	
Point		Cylinder	⬭	0.06		0.35		Median plane	
Profile of a line	✓	Parallelepiped	▱	0.07		0.4		Plane	
Surface		Sphere	◑	0.08	✓	0.45		With respect to datum plane A and datum plane B	✓
				0.09		0.5			

Example: Prs 1

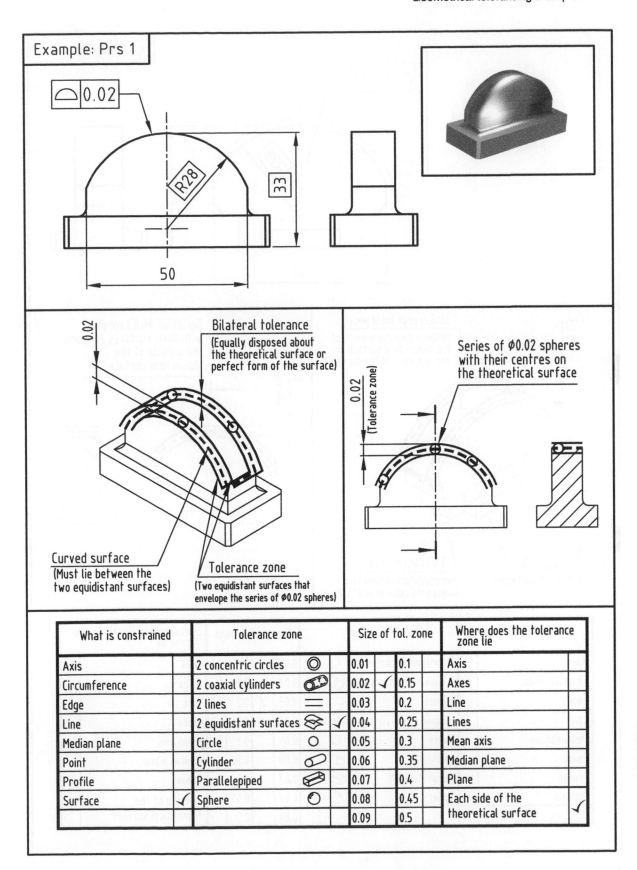

Bilateral tolerance
(Equally disposed about
the theoretical surface or
perfect form of the surface)

Curved surface
(Must lie between the
two equidistant surfaces)

Tolerance zone
(Two equidistant surfaces that
envelope the series of ⌀0.02 spheres)

Series of ⌀0.02 spheres
with their centres on
the theoretical surface

0.02 (Tolerance zone)

What is constrained		Tolerance zone		Size of tol. zone			Where does the tolerance zone lie	
Axis		2 concentric circles	◎	0.01		0.1	Axis	
Circumference		2 coaxial cylinders	⌀	0.02	✓	0.15	Axes	
Edge		2 lines	＝	0.03		0.2	Line	
Line		2 equidistant surfaces	◈ ✓	0.04		0.25	Lines	
Median plane		Circle	○	0.05		0.3	Mean axis	
Point		Cylinder	⌀	0.06		0.35	Median plane	
Profile		Parallelepiped	▱	0.07		0.4	Plane	
Surface	✓	Sphere	●	0.08		0.45	Each side of the theoretical surface	✓
				0.09		0.5		

PROFILE OF A SURFACE

Example: Prs 2

⌒ 0.03

R28

36

50

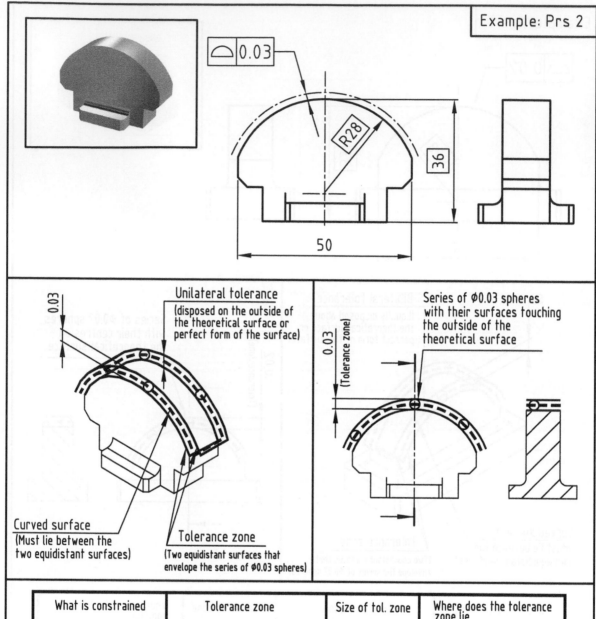

0.03

Unilateral tolerance

(disposed on the outside of the theoretical surface or perfect form of the surface)

0.03 (Tolerance zone)

Series of Ø0.03 spheres with their surfaces touching the outside of the theoretical surface

Curved surface
(Must lie between the two equidistant surfaces)

Tolerance zone
(Two equidistant surfaces that envelope the series of Ø0.03 spheres)

What is constrained		Tolerance zone		Size of tol. zone		Where does the tolerance zone lie	
Axis		2 concentric circles	◎	0.01	0.1	Axis	
Circumference		2 coaxial cylinders	⌀	0.02	0.15	Axes	
Edge		2 lines	=	0.03 ✓	0.2	Line	
Line		2 equidistant surfaces ✓	⬚	0.04	0.25	Lines	
Median plane		Circle	○	0.05	0.3	Mean axis	
Point		Cylinder	⬭	0.06	0.35	Median plane	
Profile		Parallelepiped	▱	0.07	0.4	Plane	
Surface ✓		Sphere	⬭	0.08	0.45	Outside of the theoretical surface	✓
				0.09	0.5		

PROFILE OF A SURFACE

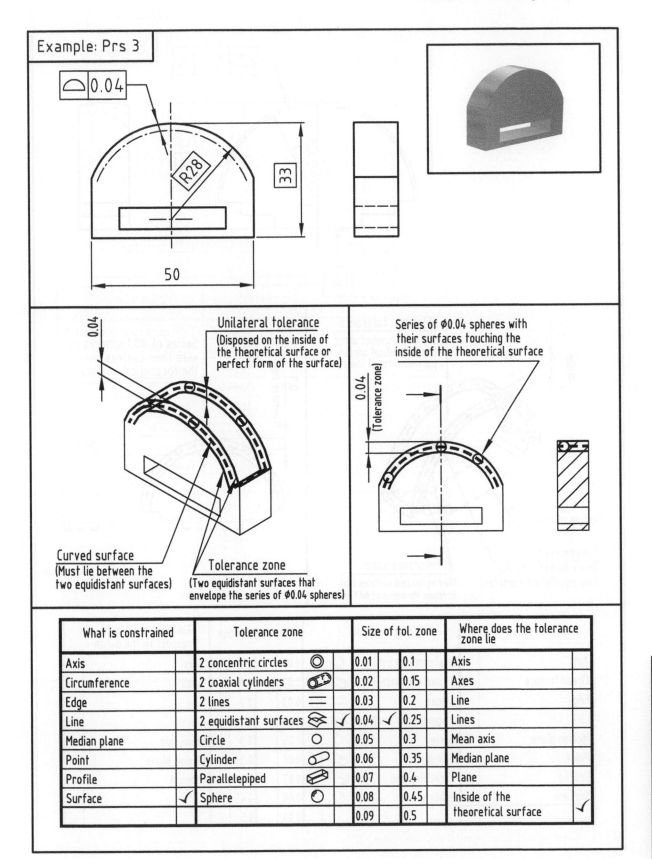

Example: Prs 3

⌒ 0.04

R28

33

50

Unilateral tolerance
(Disposed on the inside of the theoretical surface or perfect form of the surface)

0.04

Series of ⌀0.04 spheres with their surfaces touching the inside of the theoretical surface

0.04 (Tolerance zone)

Curved surface
(Must lie between the two equidistant surfaces)

Tolerance zone
(Two equidistant surfaces that envelope the series of ⌀0.04 spheres)

What is constrained		Tolerance zone		Size of tol. zone			Where does the tolerance zone lie	
Axis		2 concentric circles	◎	0.01		0.1	Axis	
Circumference		2 coaxial cylinders	⌀	0.02		0.15	Axes	
Edge		2 lines	=	0.03		0.2	Line	
Line		2 equidistant surfaces	⬥ ✓	0.04 ✓		0.25	Lines	
Median plane		Circle	○	0.05		0.3	Mean axis	
Point		Cylinder	⌀	0.06		0.35	Median plane	
Profile		Parallelepiped	▱	0.07		0.4	Plane	
Surface ✓		Sphere	○	0.08		0.45	Inside of the theoretical surface	✓
				0.09		0.5		

PROFILE OF A SURFACE

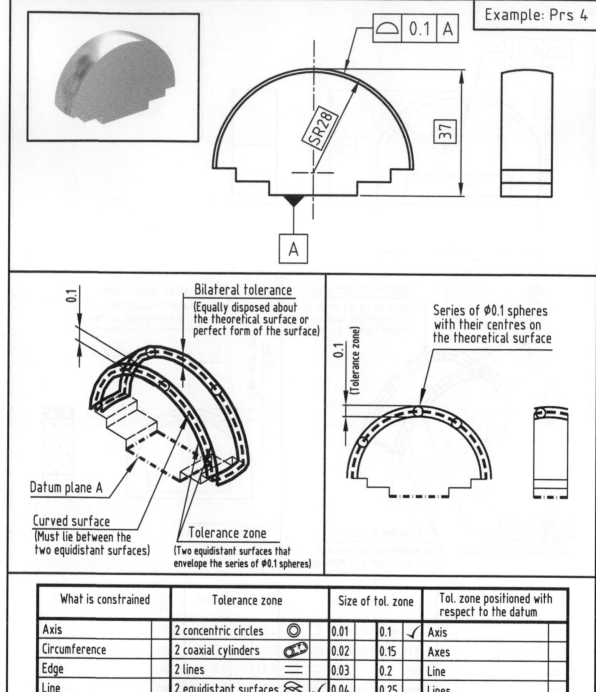

Example: Prs 4

⌓ 0.1 A

SR28

37

A

Bilateral tolerance
(Equally disposed about
the theoretical surface or
perfect form of the surface)

0.1

Datum plane A

Curved surface
(Must lie between the
two equidistant surfaces)

Tolerance zone
(Two equidistant surfaces that
envelope the series of Ø0.1 spheres)

Series of Ø0.1 spheres
with their centres on
the theoretical surface

0.1
(Tolerance zone)

What is constrained		Tolerance zone		Size of tol. zone			Tol. zone positioned with respect to the datum	
Axis		2 concentric circles	◎	0.01	0.1	✓	Axis	
Circumference		2 coaxial cylinders		0.02	0.15		Axes	
Edge		2 lines	═	0.03	0.2		Line	
Line		2 equidistant surfaces	✓	0.04	0.25		Lines	
Median plane		Circle	○	0.05	0.3		Mean axis	
Point		Cylinder	⬭	0.06	0.35		Median plane	
Profile		Parallelepiped		0.07	0.4		Plane A	✓
Surface	✓	Sphere	○	0.08	0.45			
				0.09	0.5			

Example: Rou 1

Engraved line

◯ | 0.08

Ø28

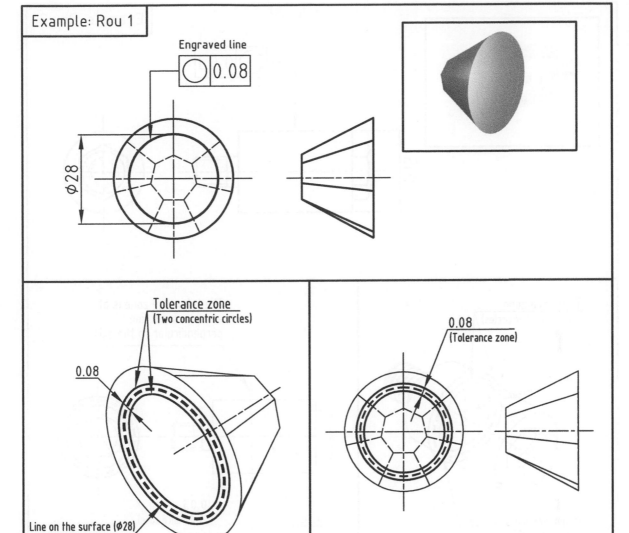

Tolerance zone
(Two concentric circles)

0.08

Line on the surface (Ø28)
(Must lie between the
two concentric circles)

0.08
(Tolerance zone)

What is constrained		Tolerance zone		Size of tol. zone		Where does the tolerance zone lie	
Axis		2 concentric circles	◎ ✓	0.01	0.1	Axis	
Circumference		2 coaxial cylinders	⌀	0.02	0.15	Axes	
Edge		2 lines	=	0.03	0.2	Line	
Line	✓	2 parallel planes	⬦	0.04	0.25	Lines	
Median plane		Circle	○	0.05	0.3	Mean axis	
Point		Cylinder	⬭	0.06	0.35	Median plane	
Profile		Parallelepiped	▱	0.07	0.4	Plane	
Surface		Sphere	◓	0.08 ✓	0.45		
				0.09	0.5	On the surface	✓

ROUNDNESS

Example: Rou 2

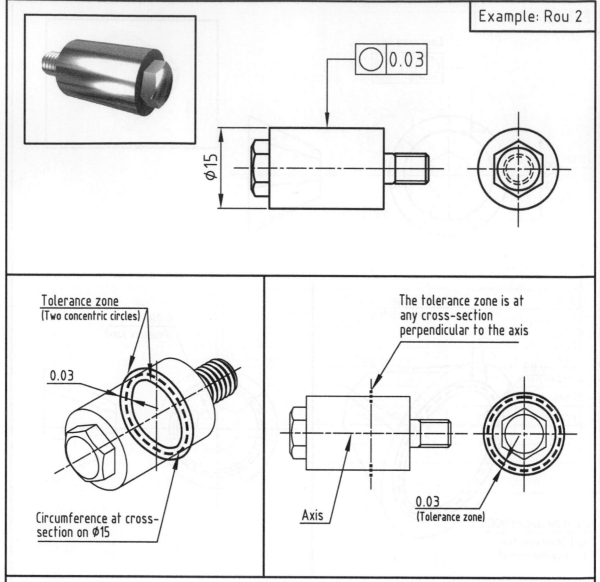

$\bigcirc$ 0.03

Ø15

Tolerance zone
(Two concentric circles)

0.03

Circumference at cross-section on Ø15

The tolerance zone is at any cross-section perpendicular to the axis

Axis

0.03
(Tolerance zone)

What is constrained	Tolerance zone		Size of tol. zone			Where does the tolerance zone lie	
Axis	2 concentric circles	◎ ✓	0.01	0.1		Axis	
Circumference	2 coaxial cylinders	⬭	0.02	0.15		Axes	
Edge	2 lines	=	0.03 ✓	0.2		Line	
Line	2 parallel planes	⬥	0.04	0.25		Lines	
Median plane	Circle	○	0.05	0.3		Mean axis	
Point	Cylinder	⬭	0.06	0.35		Median plane	
Profile	Parallelepiped	⬭	0.07	0.4		Plane	
Surface	Sphere	◐	0.08	0.45			
Circumference at cross-section ✓			0.09	0.5		In the cross-section plane	✓

Example: Rou 3

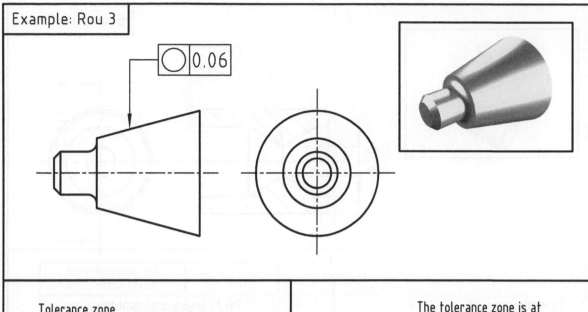

○ 0.06

Tolerance zone
(Two concentric circles)

0.06

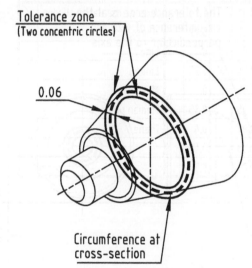

Circumference at
cross-section

The tolerance zone is at
any cross-section
perpendicular to the axis

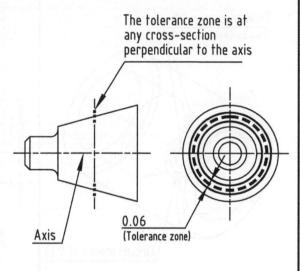

Axis

0.06
(Tolerance zone)

What is constrained	Tolerance zone			Size of tol. zone		Where does the tolerance zone lie	
Axis	2 concentric circles	◎	✓	0.01	0.1	Axis	
Circumference	2 coaxial cylinders			0.02	0.15	Axes	
Edge	2 lines	=		0.03	0.2	Line	
Line	2 parallel planes	⬥		0.04	0.25	Lines	
Median plane	Circle	○		0.05	0.3	Mean axis	
Point	Cylinder	⬭		0.06	✓ 0.35	Median plane	
Profile	Parallelepiped	▱		0.07	0.4	Plane	
Surface	Sphere	○		0.08	0.45		
Circumference at cross-section	✓			0.09	0.5	In the cross-section plane	✓

ROUNDNESS

Example: Rou 4

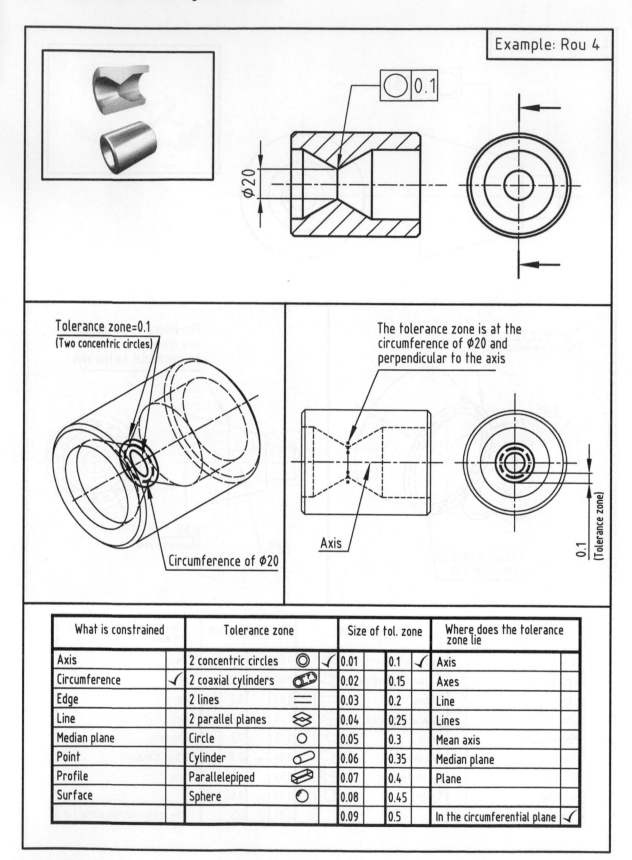

$\boxed{\bigcirc}\ \boxed{0.1}$

Ø20

Tolerance zone=0.1
(Two concentric circles)

Circumference of Ø20

The tolerance zone is at the circumference of Ø20 and perpendicular to the axis

Axis

0.1 (Tolerance zone)

What is constrained		Tolerance zone			Size of tol. zone		Where does the tolerance zone lie	
Axis		2 concentric circles	◎	✓	0.01	0.1 ✓	Axis	
Circumference	✓	2 coaxial cylinders	⌀		0.02	0.15	Axes	
Edge		2 lines	=		0.03	0.2	Line	
Line		2 parallel planes	⬦		0.04	0.25	Lines	
Median plane		Circle	○		0.05	0.3	Mean axis	
Point		Cylinder	⌀		0.06	0.35	Median plane	
Profile		Parallelepiped	⬦		0.07	0.4	Plane	
Surface		Sphere	○		0.08	0.45		
					0.09	0.5	In the circumferential plane	✓

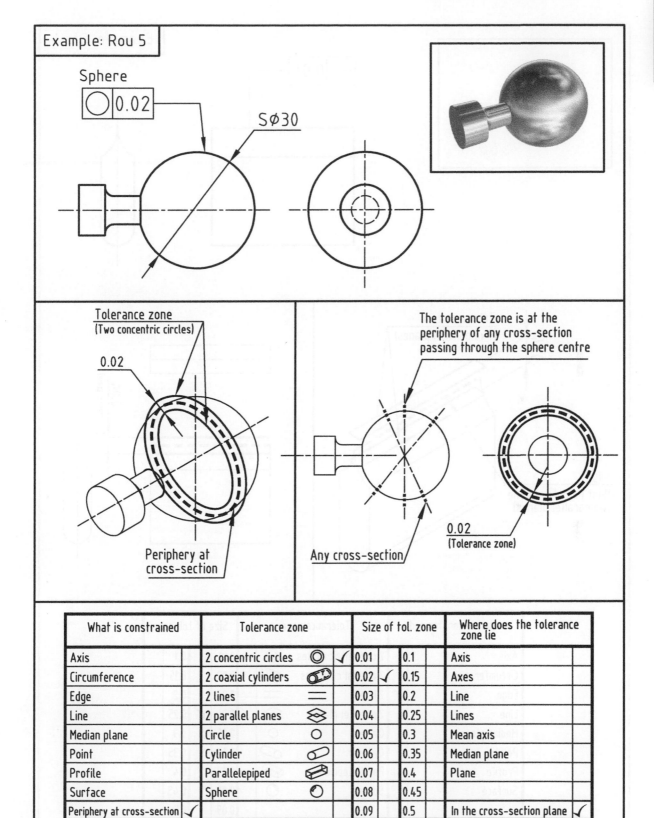

Example: Rou 5

Sphere

⬭ 0.02

SØ30

Tolerance zone (Two concentric circles)

0.02

Periphery at cross-section

The tolerance zone is at the periphery of any cross-section passing through the sphere centre

Any cross-section

0.02 (Tolerance zone)

What is constrained	Tolerance zone			Size of tol. zone		Where does the tolerance zone lie	
Axis	2 concentric circles	◎	✓	0.01	0.1	Axis	
Circumference	2 coaxial cylinders	⬭	✓	0.02 ✓	0.15	Axes	
Edge	2 lines	═		0.03	0.2	Line	
Line	2 parallel planes	⬙		0.04	0.25	Lines	
Median plane	Circle	○		0.05	0.3	Mean axis	
Point	Cylinder	⬭		0.06	0.35	Median plane	
Profile	Parallelepiped	⬙		0.07	0.4	Plane	
Surface	Sphere	◓		0.08	0.45		
Periphery at cross-section ✓				0.09	0.5	In the cross-section plane ✓	

STRAIGHTNESS

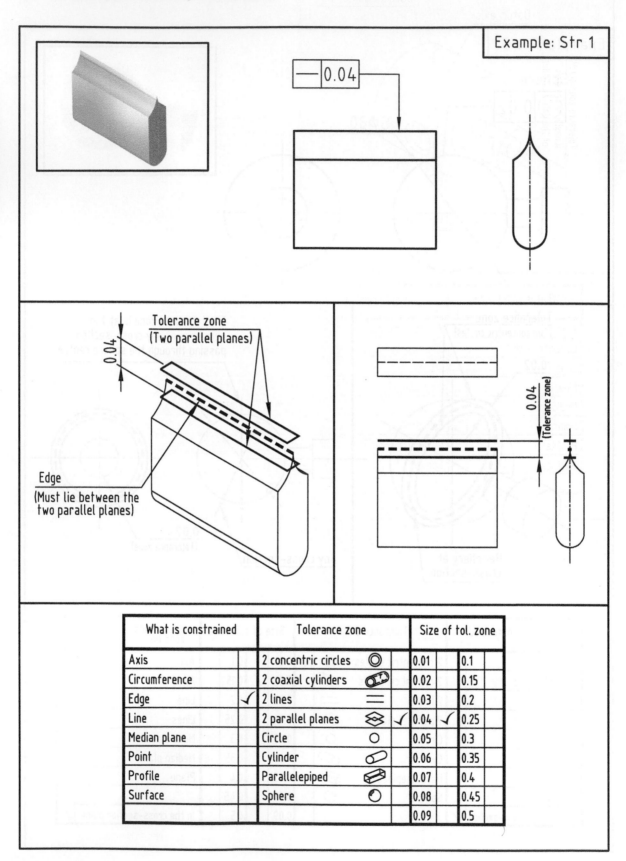

Example: Str 1

Tolerance zone
(Two parallel planes)

0.04

Edge
(Must lie between the
two parallel planes)

0.04
(Tolerance zone)

What is constrained		Tolerance zone		Size of tol. zone		
Axis		2 concentric circles	◎	0.01		0.1
Circumference		2 coaxial cylinders	⌀	0.02		0.15
Edge	✓	2 lines	═	0.03		0.2
Line		2 parallel planes	⬦ ✓	0.04 ✓		0.25
Median plane		Circle	○	0.05		0.3
Point		Cylinder	⬭	0.06		0.35
Profile		Parallelepiped	▱	0.07		0.4
Surface		Sphere	◑	0.08		0.45
				0.09		0.5

Example: Str 2

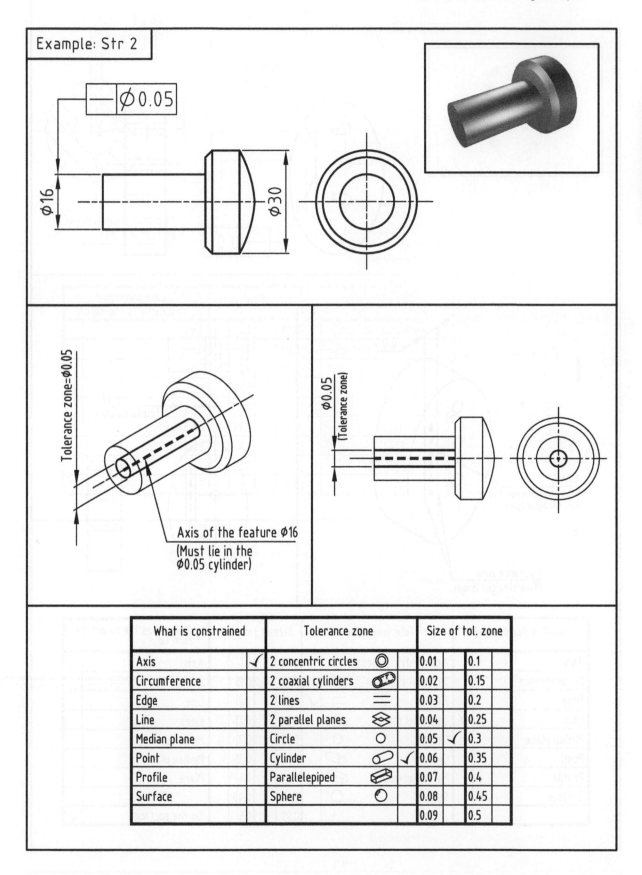

Axis of the feature Ø16
(Must lie in the
Ø0.05 cylinder)

STRAIGHTNESS

What is constrained		Tolerance zone		Size of tol. zone			
Axis	✓	2 concentric circles	◎	0.01		0.1	
Circumference		2 coaxial cylinders	⌀	0.02		0.15	
Edge		2 lines	=	0.03		0.2	
Line		2 parallel planes	⬨	0.04		0.25	
Median plane		Circle	○	0.05	✓	0.3	
Point		Cylinder	⌀	✓	0.06	0.35	
Profile		Parallelepiped	▱	0.07		0.4	
Surface		Sphere	⬭	0.08		0.45	
				0.09		0.5	

STRAIGHTNESS

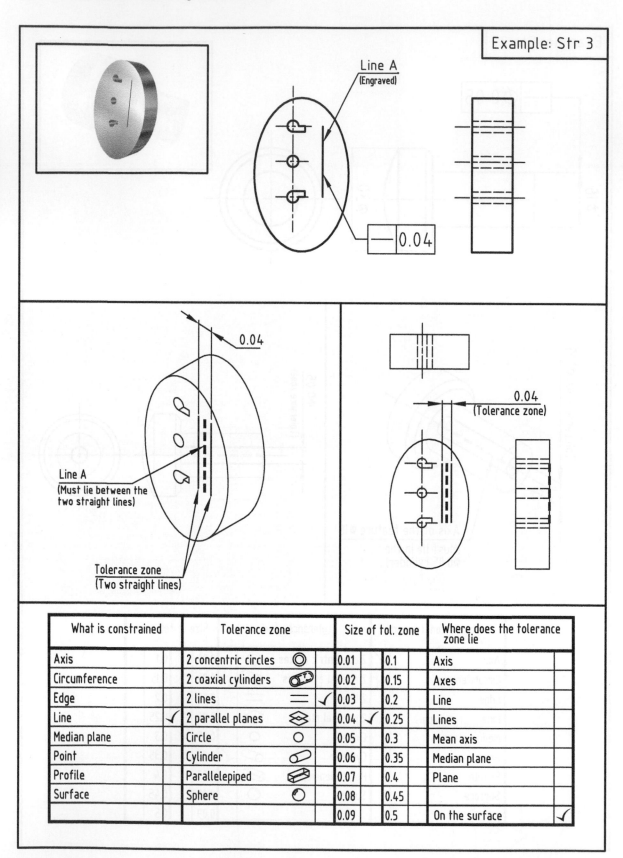

Example: Str 3

Line A (Engraved)

─ | 0.04

0.04

Line A
(Must lie between the two straight lines)

Tolerance zone
(Two straight lines)

0.04
(Tolerance zone)

What is constrained		Tolerance zone			Size of tol. zone		Where does the tolerance zone lie	
Axis		2 concentric circles	◎		0.01	0.1	Axis	
Circumference		2 coaxial cylinders	⬭		0.02	0.15	Axes	
Edge		2 lines	═	✓	0.03	0.2	Line	
Line	✓	2 parallel planes	⬖		0.04	✓ 0.25	Lines	
Median plane		Circle	○		0.05	0.3	Mean axis	
Point		Cylinder	⌬		0.06	0.35	Median plane	
Profile		Parallelepiped	▱		0.07	0.4	Plane	
Surface		Sphere	⬯		0.08	0.45		
					0.09	0.5	On the surface	✓

STRAIGHTNESS

Example: Str 4

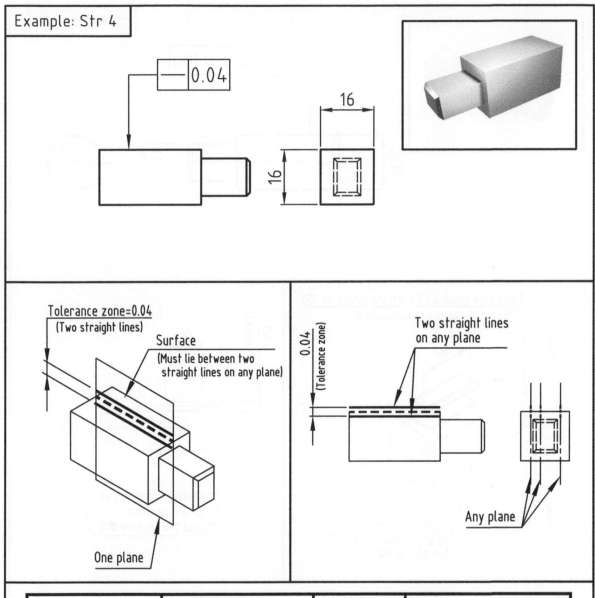

What is constrained		Tolerance zone		Size of tol. zone		Where does the tolerance zone lie	
Axis		2 concentric circles ◎		0.01	0.1	Axis	
Circumference		2 coaxial cylinders ⌾		0.02	0.15	Axes	
Edge		2 lines ═	✓	0.03	0.2	Line	
Line		2 parallel planes ⬙	✓	0.04 ✓	0.25	Lines	
Median plane		Circle ○		0.05	0.3	Mean axis	
Point		Cylinder ⌭		0.06	0.35	Median plane	
Profile		Parallelepiped ⬙		0.07	0.4	On any plane	✓
Surface	✓	Sphere ⬭		0.08	0.45		
				0.09	0.5		

STRAIGHTNESS

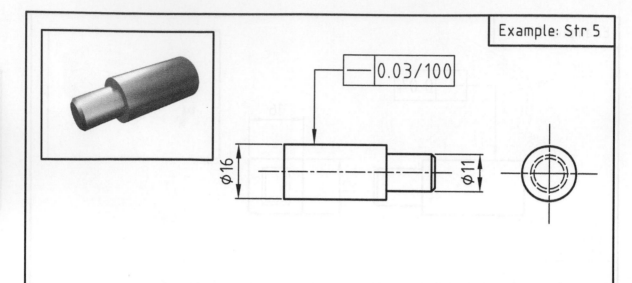

Example: Str 5

0.03/100

Ø16

Ø11

Tolerance zone=0.03 over any length of 100
(Two parallel planes at any position)

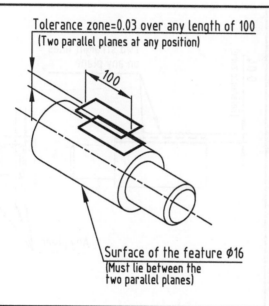

100

Surface of the feature Ø16
(Must lie between the
two parallel planes)

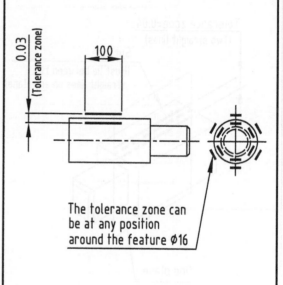

0.03
(Tolerance zone)

100

The tolerance zone can
be at any position
around the feature Ø16

What is constrained		Tolerance zone		Size of tol. zone			Where does the tolerance zone lie	
Axis		2 concentric circles	◎	0.01		0.1	Axis	
Circumference		2 coaxial cylinders	⬭	0.02		0.15	Axes	
Edge		2 lines	═	0.03	✓	0.2	Line	
Line		2 parallel planes	⬖ ✓	0.04		0.25	Lines	
Median plane		Circle	○	0.05		0.3	Mean axis	
Point		Cylinder	�	0.06		0.35	Median plane	
Profile		Parallelepiped	▱	0.07		0.4	Plane	
Surface	✓	Sphere	⬯	0.08		0.45		
				0.09		0.5	All positions around feature Ø16	✓

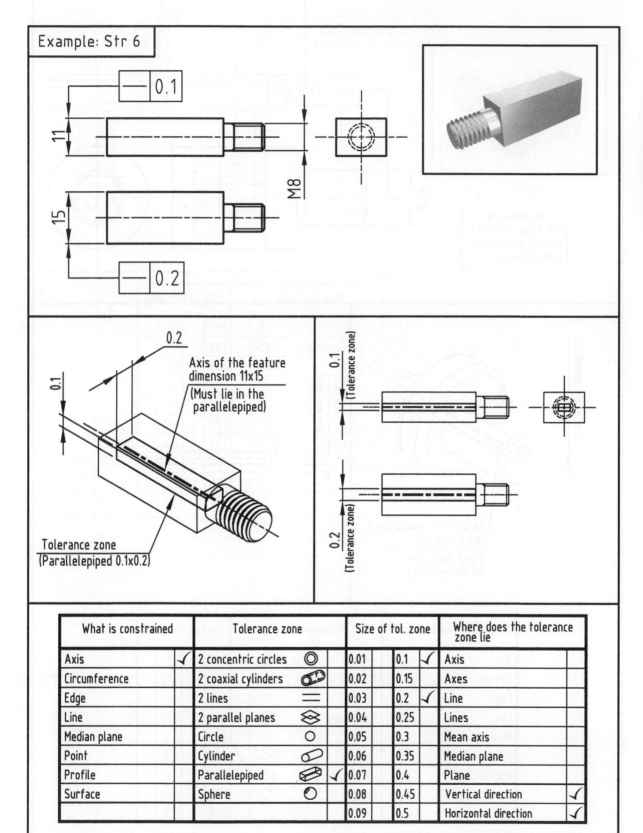

Example: Str 6

0.1

11

15

M8

0.2

0.2

0.1

Axis of the feature dimension 11x15
(Must lie in the parallelepiped)

0.1 (Tolerance zone)

0.2 (Tolerance zone)

Tolerance zone
(Parallelepiped 0.1x0.2)

What is constrained		Tolerance zone		Size of tol. zone			Where does the tolerance zone lie	
Axis	✓	2 concentric circles	◎	0.01	0.1	✓	Axis	
Circumference		2 coaxial cylinders	⌀	0.02	0.15		Axes	
Edge		2 lines	═	0.03	0.2	✓	Line	
Line		2 parallel planes	◈	0.04	0.25		Lines	
Median plane		Circle	○	0.05	0.3		Mean axis	
Point		Cylinder	⌀	0.06	0.35		Median plane	
Profile		Parallelepiped	▱	✓ 0.07	0.4		Plane	
Surface		Sphere	○	0.08	0.45		Vertical direction	✓
				0.09	0.5		Horizontal direction	✓

STRAIGHTNESS

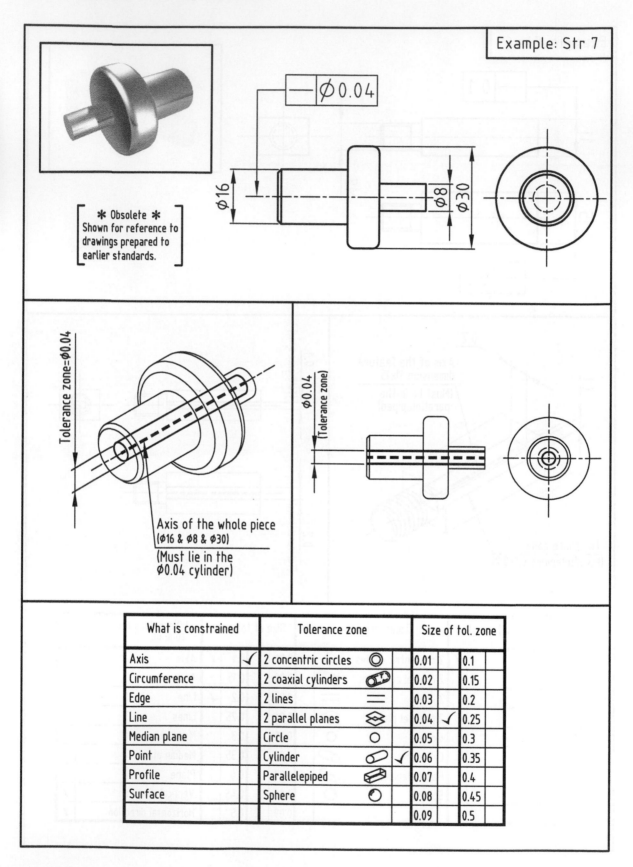

Example: Str 7

∅0.04

∅16

∅8

∅30

∗ Obsolete ∗
Shown for reference to
drawings prepared to
earlier standards.

Tolerance zone=∅0.04

∅0.04
(Tolerance zone)

Axis of the whole piece
(∅16 & ∅8 & ∅30)

(Must lie in the
∅0.04 cylinder)

What is constrained		Tolerance zone		Size of tol. zone		
Axis	✓	2 concentric circles	◎	0.01	0.1	
Circumference		2 coaxial cylinders		0.02	0.15	
Edge		2 lines	=	0.03	0.2	
Line		2 parallel planes	◈	0.04 ✓	0.25	
Median plane		Circle	○	0.05	0.3	
Point		Cylinder	⬭ ✓	0.06	0.35	
Profile		Parallelepiped	▱	0.07	0.4	
Surface		Sphere	◯	0.08	0.45	
				0.09	0.5	

STRAIGHTNESS

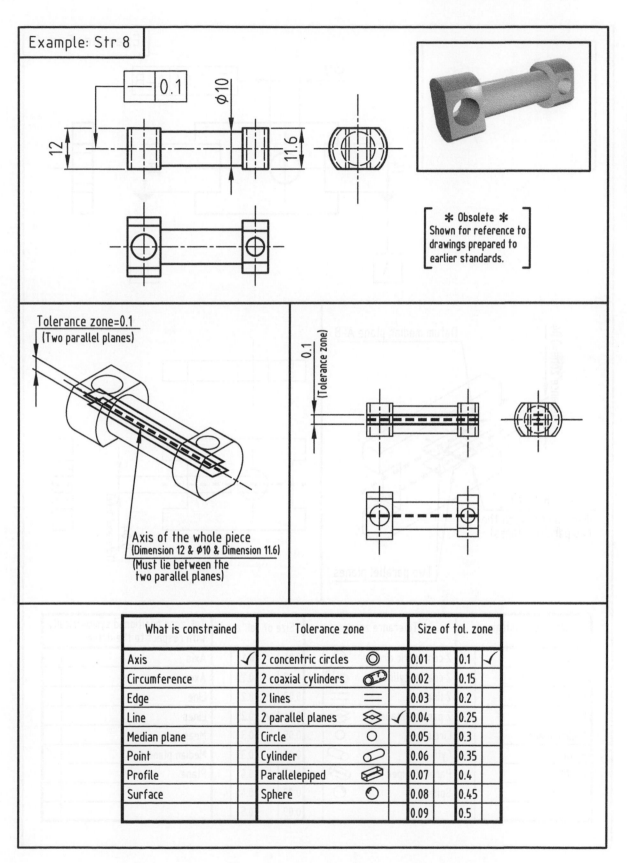

Example: Str 8

0.1

⌀10

12

11.6

STRAIGHTNESS

Tolerance zone=0.1
(Two parallel planes)

Axis of the whole piece
(Dimension 12 & ⌀10 & Dimension 11.6)
(Must lie between the
two parallel planes)

0.1
(Tolerance zone)

What is constrained		Tolerance zone		Size of tol. zone			
Axis	✓	2 concentric circles	◎	0.01		0.1	✓
Circumference		2 coaxial cylinders	⌀	0.02		0.15	
Edge		2 lines	═	0.03		0.2	
Line		2 parallel planes	◇ ✓	0.04		0.25	
Median plane		Circle	○	0.05		0.3	
Point		Cylinder	⌀	0.06		0.35	
Profile		Parallelepiped	▱	0.07		0.4	
Surface		Sphere	○	0.08		0.45	
				0.09		0.5	

SYMMETRY

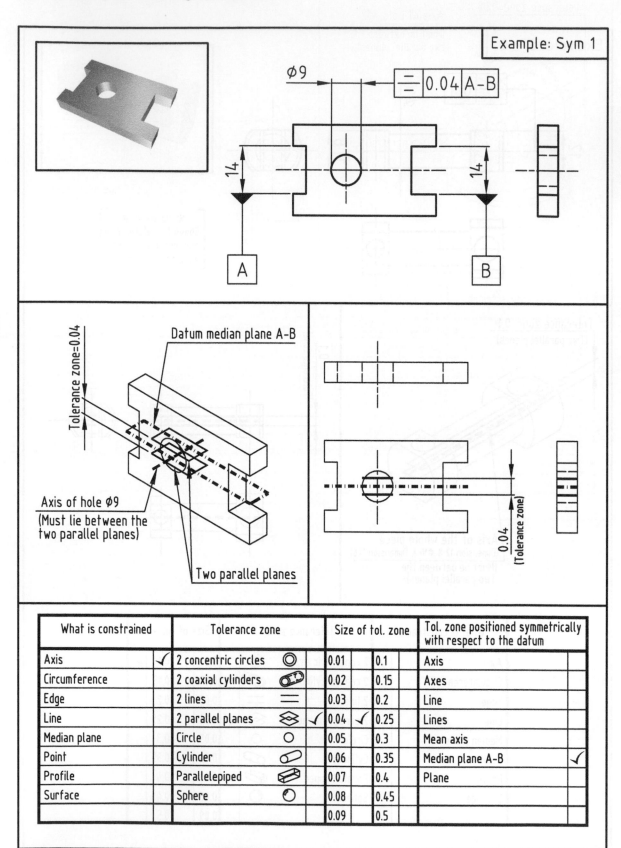

Example: Sym 1

Datum median plane A-B

Tolerance zone=0.04

Axis of hole Ø9
(Must lie between the
two parallel planes)

Two parallel planes

0.04
(Tolerance zone)

What is constrained		Tolerance zone		Size of tol. zone		Tol. zone positioned symmetrically with respect to the datum	
Axis	✓	2 concentric circles	◎	0.01	0.1	Axis	
Circumference		2 coaxial cylinders	⌀	0.02	0.15	Axes	
Edge		2 lines	═	0.03	0.2	Line	
Line		2 parallel planes	⬨ ✓	0.04 ✓	0.25	Lines	
Median plane		Circle	○	0.05	0.3	Mean axis	
Point		Cylinder	⌀	0.06	0.35	Median plane A-B	✓
Profile		Parallelepiped	▱	0.07	0.4	Plane	
Surface		Sphere	◉	0.08	0.45		
				0.09	0.5		

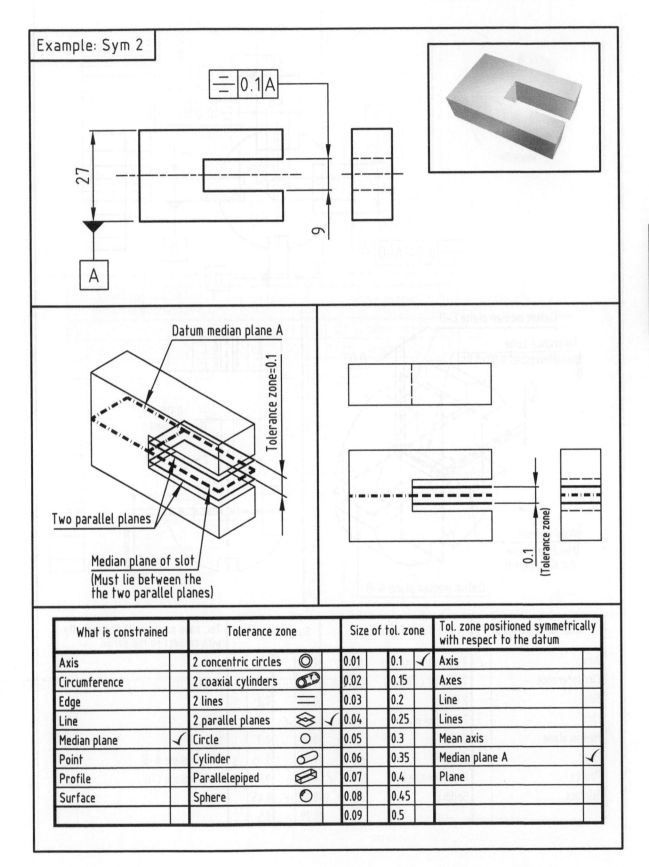

Example: Sym 2

27

9

A

Datum median plane A

Tolerance zone=0.1

Two parallel planes

Median plane of slot
(Must lie between the
the two parallel planes)

0.1
(Tolerance zone)

What is constrained		Tolerance zone		Size of tol. zone				Tol. zone positioned symmetrically with respect to the datum	
Axis		2 concentric circles	◎	0.01		0.1	✓	Axis	
Circumference		2 coaxial cylinders	⌀	0.02		0.15		Axes	
Edge		2 lines	=	0.03		0.2		Line	
Line		2 parallel planes	◈	✓ 0.04		0.25		Lines	
Median plane	✓	Circle	○	0.05		0.3		Mean axis	
Point		Cylinder	⌀	0.06		0.35		Median plane A	✓
Profile		Parallelepiped	▱	0.07		0.4		Plane	
Surface		Sphere	○	0.08		0.45			
				0.09		0.5			

SYMMETRY

SYMMETRY

Example: Sym 3

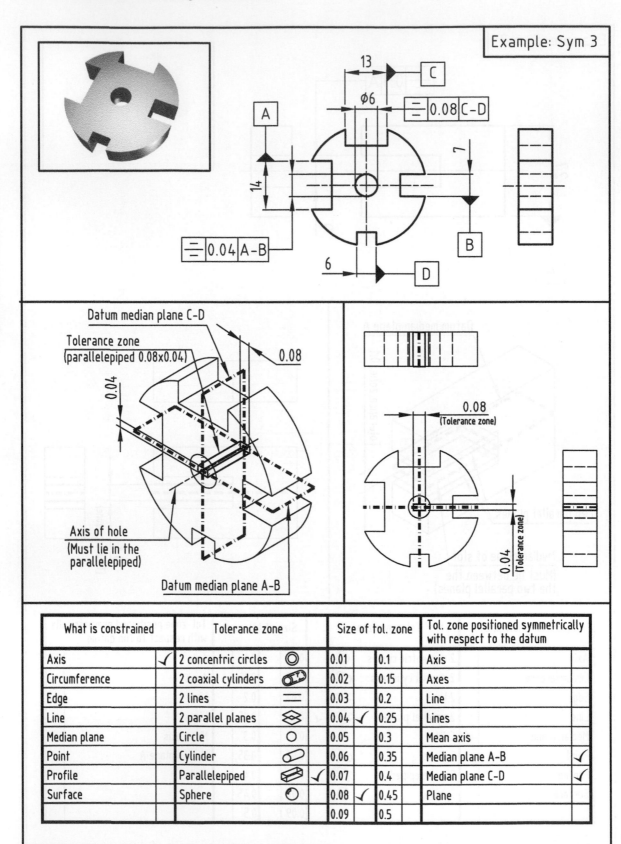

Datum median plane C–D

Tolerance zone
(parallelepiped 0.08x0.04)

Axis of hole
(Must lie in the
parallelepiped)

Datum median plane A–B

0.08
(Tolerance zone)

0.04
(Tolerance zone)

What is constrained		Tolerance zone		Size of tol. zone				Tol. zone positioned symmetrically with respect to the datum	
Axis	✓	2 concentric circles	◎	0.01		0.1		Axis	
Circumference		2 coaxial cylinders	⌀	0.02		0.15		Axes	
Edge		2 lines	═	0.03		0.2		Line	
Line		2 parallel planes	⬥	0.04	✓	0.25		Lines	
Median plane		Circle	○	0.05		0.3		Mean axis	
Point		Cylinder	⌀	0.06		0.35		Median plane A–B	✓
Profile		Parallelepiped	▱	✓ 0.07		0.4		Median plane C–D	✓
Surface		Sphere	◓	0.08	✓	0.45		Plane	
				0.09		0.5			

SYMMETRY

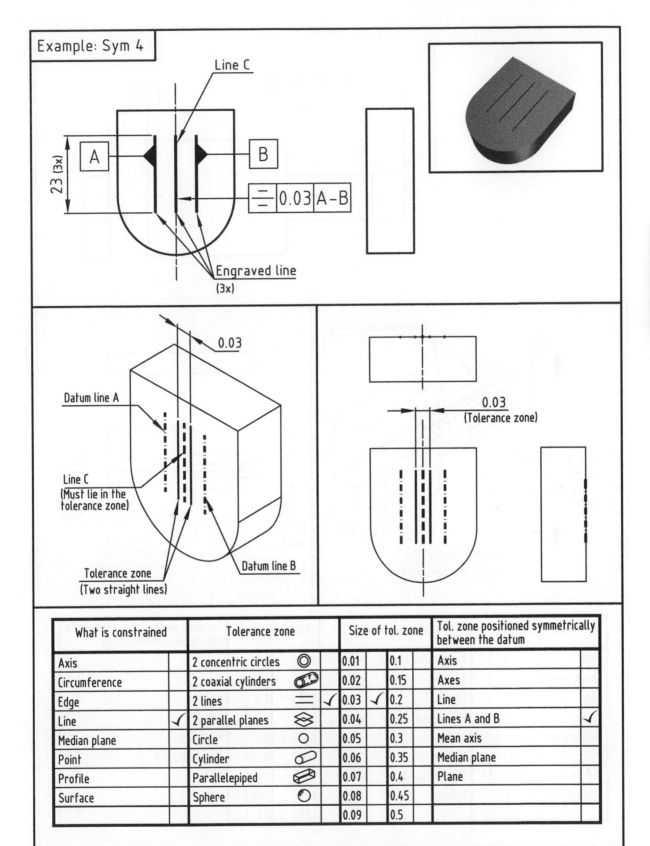

What is constrained		Tolerance zone		Size of tol. zone				Tol. zone positioned symmetrically between the datum	
Axis		2 concentric circles	⊚	0.01		0.1		Axis	
Circumference		2 coaxial cylinders	⌀	0.02		0.15		Axes	
Edge		2 lines	═	✓	0.03	✓	0.2	Line	
Line	✓	2 parallel planes	⬨	0.04		0.25		Lines A and B	✓
Median plane		Circle	○	0.05		0.3		Mean axis	
Point		Cylinder	⬭	0.06		0.35		Median plane	
Profile		Parallelepiped	▱	0.07		0.4		Plane	
Surface		Sphere	◍	0.08		0.45			
				0.09		0.5			

SYMMETRY

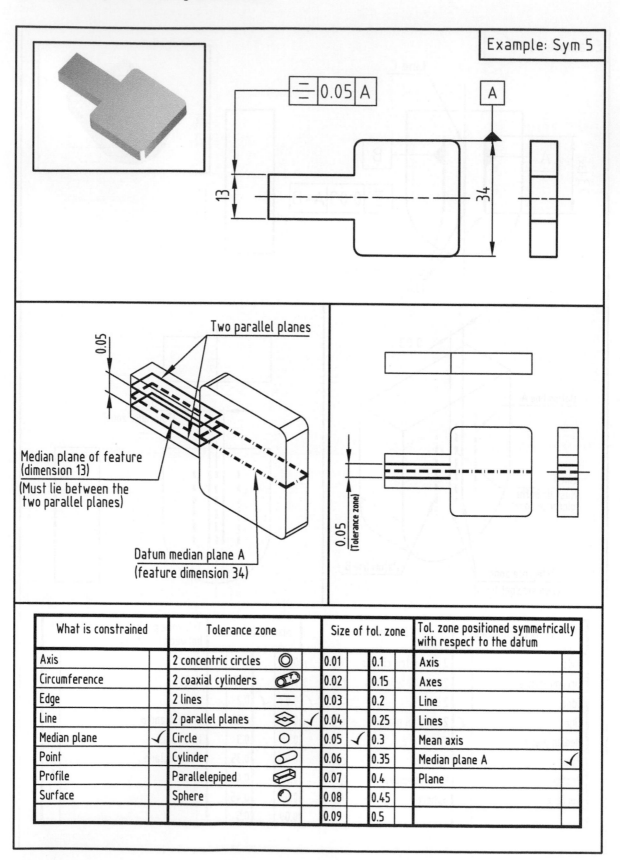

Example: Sym 5

Two parallel planes

Median plane of feature
(dimension 13)
(Must lie between the
two parallel planes)

Datum median plane A
(feature dimension 34)

What is constrained		Tolerance zone		Size of tol. zone			Tol. zone positioned symmetrically with respect to the datum	
Axis		2 concentric circles	◎	0.01		0.1	Axis	
Circumference		2 coaxial cylinders	⬭	0.02		0.15	Axes	
Edge		2 lines	=	0.03		0.2	Line	
Line		2 parallel planes	⬥ ✓	0.04		0.25	Lines	
Median plane	✓	Circle	○	0.05 ✓		0.3	Mean axis	
Point		Cylinder	⬭	0.06		0.35	Median plane A	✓
Profile		Parallelepiped	▱	0.07		0.4	Plane	
Surface		Sphere	◕	0.08		0.45		
				0.09		0.5		

Example: Sym 6

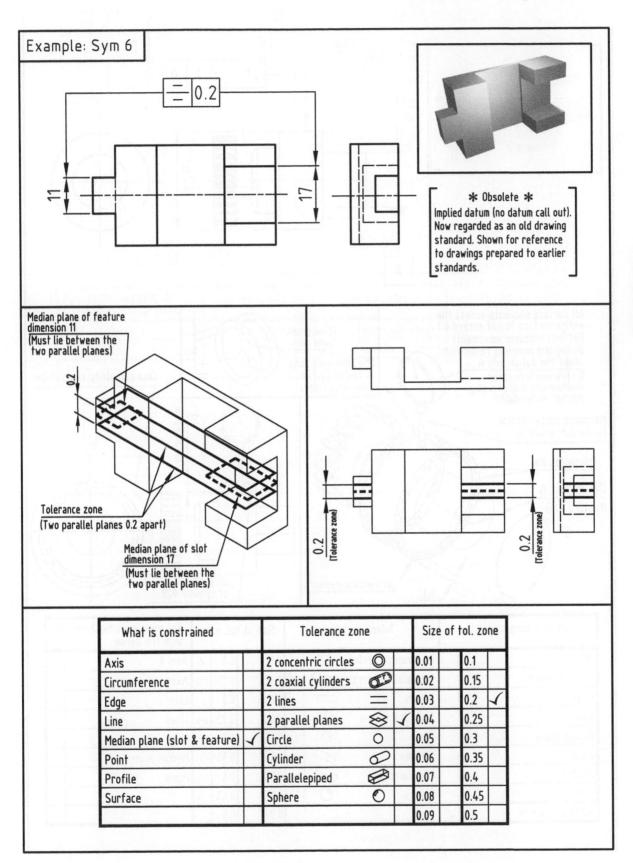

11

17

‖ 0.2

* Obsolete *
Implied datum (no datum call out).
Now regarded as an old drawing
standard. Shown for reference
to drawings prepared to earlier
standards.

SYMMETRY

Median plane of feature
dimension 11
(Must lie between the
two parallel planes)

0.2

Tolerance zone
(Two parallel planes 0.2 apart)

Median plane of slot
dimension 17
(Must lie between the
two parallel planes)

0.2 [Tolerance zone]

0.2 [Tolerance zone]

What is constrained		Tolerance zone		Size of tol. zone		
Axis		2 concentric circles	◎	0.01	0.1	
Circumference		2 coaxial cylinders	⌬	0.02	0.15	
Edge		2 lines	=	0.03	0.2	✓
Line		2 parallel planes	⬨ ✓	0.04	0.25	
Median plane (slot & feature)	✓	Circle	○	0.05	0.3	
Point		Cylinder	⌀	0.06	0.35	
Profile		Parallelepiped	▱	0.07	0.4	
Surface		Sphere	⬤	0.08	0.45	
				0.09	0.5	

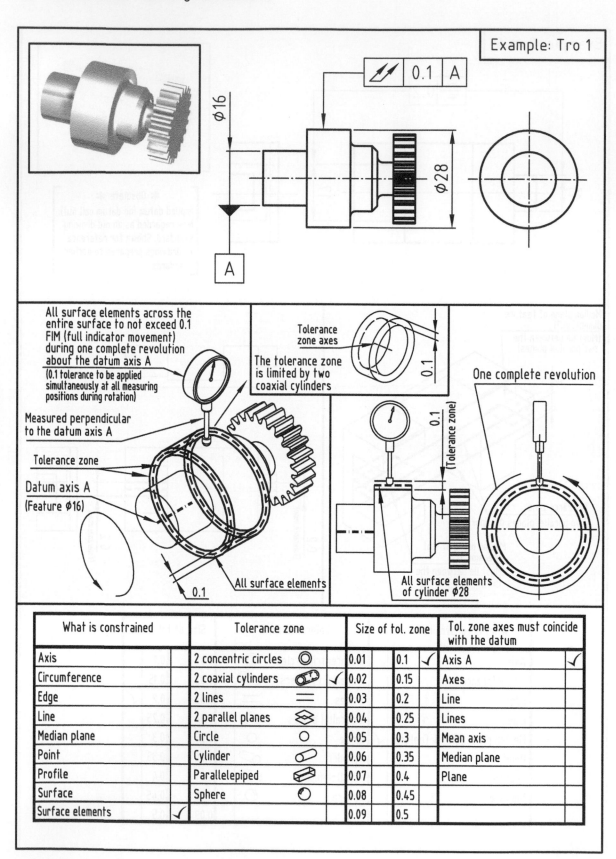

TOTAL RUN-OUT

Example: Tro 1

⌀16

⌀28

All surface elements across the entire surface to not exceed 0.1 FIM (full indicator movement) during one complete revolution about the datum axis A
(0.1 tolerance to be applied simultaneously at all measuring positions during rotation)

Measured perpendicular to the datum axis A

Tolerance zone

Datum axis A
(Feature ⌀16)

0.1

All surface elements

Tolerance zone axes

The tolerance zone is limited by two coaxial cylinders

0.1

0.1

(Tolerance zone)

One complete revolution

All surface elements of cylinder ⌀28

What is constrained		Tolerance zone		Size of tol. zone			Tol. zone axes must coincide with the datum	
Axis		2 concentric circles	◎	0.01	✓	0.1	Axis A	✓
Circumference		2 coaxial cylinders	⌯ ✓	0.02		0.15	Axes	
Edge		2 lines	=	0.03		0.2	Line	
Line		2 parallel planes	⬥	0.04		0.25	Lines	
Median plane		Circle	○	0.05		0.3	Mean axis	
Point		Cylinder	⌀	0.06		0.35	Median plane	
Profile		Parallelepiped	▱	0.07		0.4	Plane	
Surface		Sphere	◑	0.08		0.45		
Surface elements	✓			0.09		0.5		

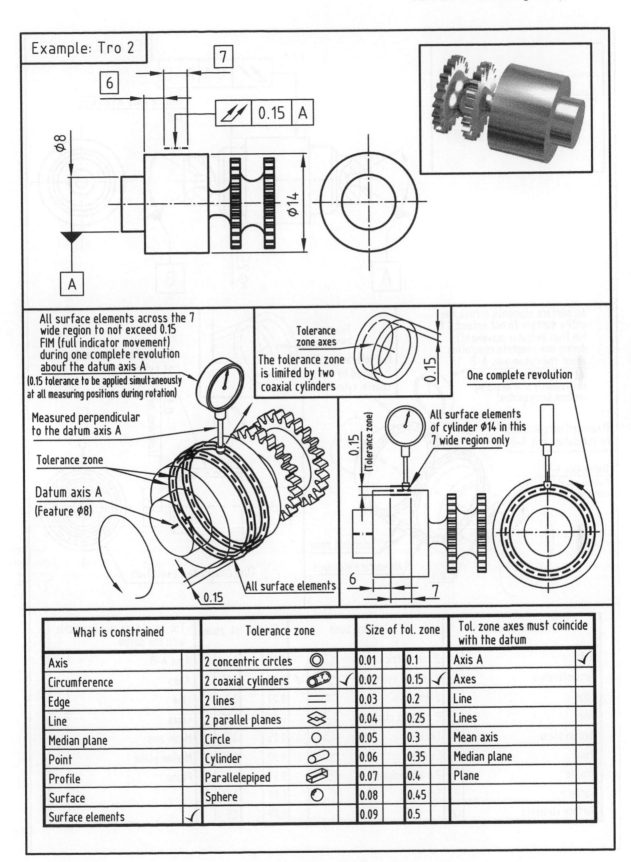

Example: Tro 2

⑥ ⑦

⟋⟋ | 0.15 | A

Ø8

Ø14

A

All surface elements across the 7 wide region to not exceed 0.15 FIM (full indicator movement) during one complete revolution about the datum axis A
(0.15 tolerance to be applied simultaneously at all measuring positions during rotation)

Measured perpendicular to the datum axis A

Tolerance zone

Datum axis A
(Feature Ø8)

All surface elements

0.15

Tolerance zone axes

The tolerance zone is limited by two coaxial cylinders

0.15

0.15 (Tolerance zone)

All surface elements of cylinder Ø14 in this 7 wide region only

One complete revolution

6   7

TOTAL RUN-OUT

What is constrained		Tolerance zone		Size of tol. zone		Tol. zone axes must coincide with the datum	
Axis		2 concentric circles	◎	0.01	0.1	Axis A	✓
Circumference		2 coaxial cylinders	⌀ ✓	0.02	0.15 ✓	Axes	
Edge		2 lines	═	0.03	0.2	Line	
Line		2 parallel planes	⬙	0.04	0.25	Lines	
Median plane		Circle	○	0.05	0.3	Mean axis	
Point		Cylinder	⌀	0.06	0.35	Median plane	
Profile		Parallelepiped	⬙	0.07	0.4	Plane	
Surface		Sphere	◓	0.08	0.45		
Surface elements	✓			0.09	0.5		

Example: Tro 3

Centre hole A1.6x3.35 (2x)

⌀25

A

B

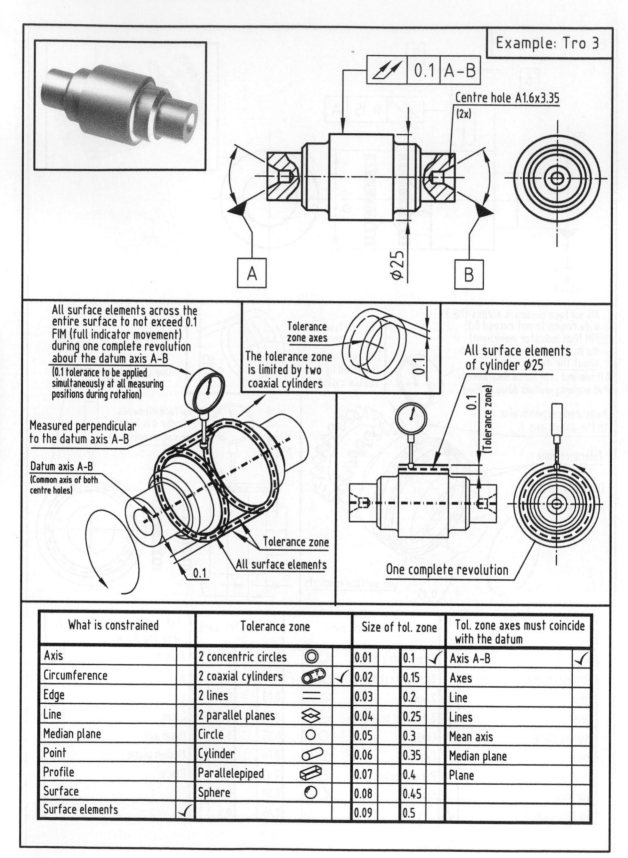

TOTAL RUN-OUT

All surface elements across the entire surface to not exceed 0.1 FIM (full indicator movement) during one complete revolution about the datum axis A-B
(0.1 tolerance to be applied simultaneously at all measuring positions during rotation)

Tolerance zone axes

The tolerance zone is limited by two coaxial cylinders

Measured perpendicular to the datum axis A-B

Datum axis A-B
(Common axis of both centre holes)

All surface elements of cylinder ⌀25

(Tolerance zone)

Tolerance zone

All surface elements

0.1

One complete revolution

What is constrained	Tolerance zone		Size of tol. zone			Tol. zone axes must coincide with the datum	
Axis	2 concentric circles	◎	0.01	0.1	✓	Axis A-B	✓
Circumference	2 coaxial cylinders	⬭ ✓	0.02	0.15		Axes	
Edge	2 lines	=	0.03	0.2		Line	
Line	2 parallel planes	⬦	0.04	0.25		Lines	
Median plane	Circle	○	0.05	0.3		Mean axis	
Point	Cylinder	⬭	0.06	0.35		Median plane	
Profile	Parallelepiped	▱	0.07	0.4		Plane	
Surface	Sphere	◐	0.08	0.45			
Surface elements	✓		0.09	0.5			

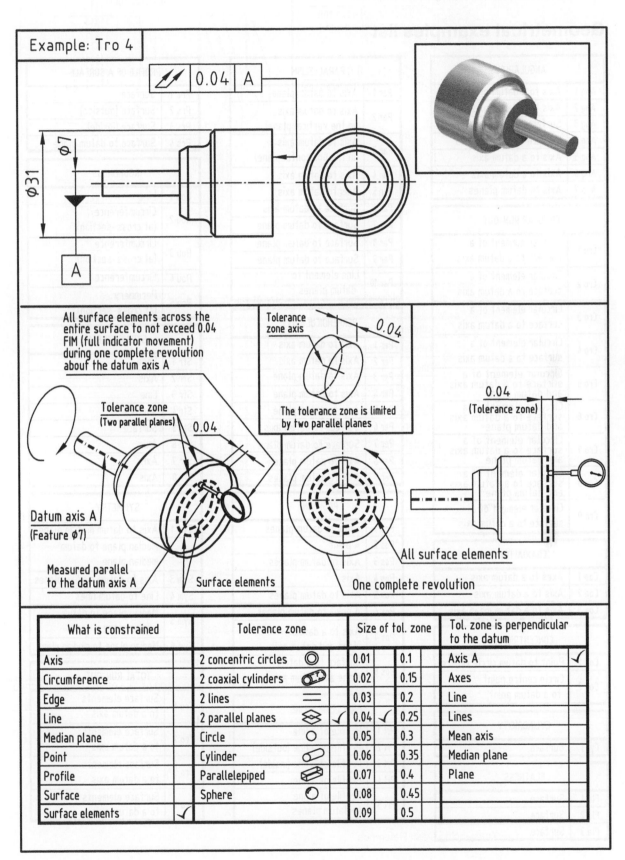

Example: Tro 4

⟋⟋ | 0.04 | A

Ø31   Ø7

A

All surface elements across the entire surface to not exceed 0.04 FIM (full indicator movement) during one complete revolution about the datum axis A

Tolerance zone
(Two parallel planes)   0.04

Datum axis A
(Feature Ø7)

Measured parallel
to the datum axis A   Surface elements

Tolerance
zone axis   0.04

The tolerance zone is limited
by two parallel planes

0.04
(Tolerance zone)

All surface elements

One complete revolution

What is constrained	Tolerance zone		Size of tol. zone		Tol. zone is perpendicular to the datum	
Axis	2 concentric circles	◎	0.01	0.1	Axis A	✓
Circumference	2 coaxial cylinders	⌀	0.02	0.15	Axes	
Edge	2 lines	=	0.03	0.2	Line	
Line	2 parallel planes ✓	⬙	0.04 ✓	0.25	Lines	
Median plane	Circle	○	0.05	0.3	Mean axis	
Point	Cylinder	⌾	0.06	0.35	Median plane	
Profile	Parallelepiped	▱	0.07	0.4	Plane	
Surface	Sphere	◔	0.08	0.45		
Surface elements	✓		0.09	0.5		

TOTAL RUN-OUT

Geometrical examples list

ANGULARITY	
Ang 1	Axis to a datum plane
Ang 2	Axis to a datum axis
Ang 3	Surface to a datum plane
Ang 4	Surface to a datum axis
Ang 5	Axis to a datum axis
Ang 6	Axis to a datum axis
Ang 7	Axis to datum planes

CIRCULAR RUN-OUT	
Cro 1	Circular element of a surface to a datum axis
Cro 2	Circular element of a surface to a datum axis
Cro 3	Circular element of a surface to a datum axis
Cro 4	Circular element of a surface to a datum axis
Cro 5	Circular element of a surface to a datum axis and datum plane
Cro 6	Circular element of a surface to a datum axis and datum plane
Cro 7	Circular element of a surface to a datum axis and datum plane
Cro 8	Circular element of a surface to a datum axis and datum plane
Cro 9	Circular element of a surface to a datum axis

COAXIALITY	
Coa 1	Axes to a datum axis
Coa 2	Axis to a datum axis
Coa 3	Axis to a datum mean axis

CONCENTRICITY	
Con 1	Point to datum circle centre
Con 2	Circle centre point to a datum point

CYLINDRICITY	
Cyl 1	Surface

FLATNESS	
Fla 1	Surface
Fla 2	Surface
Fla 3	Surface

PARALLELISM	
Par 1	Axis to datum plane
Par 2	Axis to datum axis (in the vertical plane)
Par 3	Axis to datum axis (in the horizontal plane)
Par 4	Axis to datum axis
Par 5	Axis to datum axis
Par 6	Surface to datum axis
Par 7	Surface to datum plane
Par 8	Surface to datum plane
Par 9	Surface to datum plane
Par 10	Line element to datum planes

PERPENDICULARITY	
Per 1	Axis to datum axis
Per 2	Axis to datum axis
Per 3	Axis to datum plane
Per 4	Axis to datum plane
Per 5	Axis to datum plane
Per 6	Surface to datum axis
Per 7	Surface to datum plane
Per 8	Axis to datum planes
Per 9	Axis to datum planes

POSITION	
Pos 1	Lines to datum planes
Pos 2	A point
Pos 3	Axis to datum planes
Pos 4	Axis
Pos 5	Axis to datum planes
Pos 6	A point to datum planes
Pos 7	Axis to a datum plane and datum axis
Pos 8	Surface to a datum plane and datum axis

PROFILE OF A LINE	
Prl 1	Profile of a line
Prl 2	Profile of a line (outside)
Prl 3	Profile of a line (inside)
Prl 4	Profile of a line
Prl 5	Profile of a line to datum planes

PROFILE OF A SURFACE	
Prs 1	Surface
Prs 2	Surface (outside)
Prs 3	Surface (inside)
Prs 4	Surface to datum plane

ROUNDNESS	
Rou 1	Line
Rou 2	Circumference (at cross-section)
Rou 3	Circumference (at cross-section)
Rou 4	Circumference
Rou 5	Periphery (at cross-section)

STRAIGHTNESS	
Str 1	Edge
Str 2	Axis
Str 3	Line
Str 4	Surface
Str 5	Surface
Str 6	Axis
Str 7	Axis
Str 8	Axis

SYMMETRY	
Sym 1	Axis to datum median plane
Sym 2	Median plane to datum median plane
Sym 3	Axis to datum median planes
Sym 4	Line to datum lines
Sym 5	Median plane to datum median plane
Sym 6	Median plane to median plane

TOTAL RUN-OUT	
Tro 1	Surface elements to a datum axis
Tro 2	Surface elements to a datum axis
Tro 3	Surface elements to a datum axis
Tro 4	Surface elements to a datum axis

PART 4

Quick reference tables

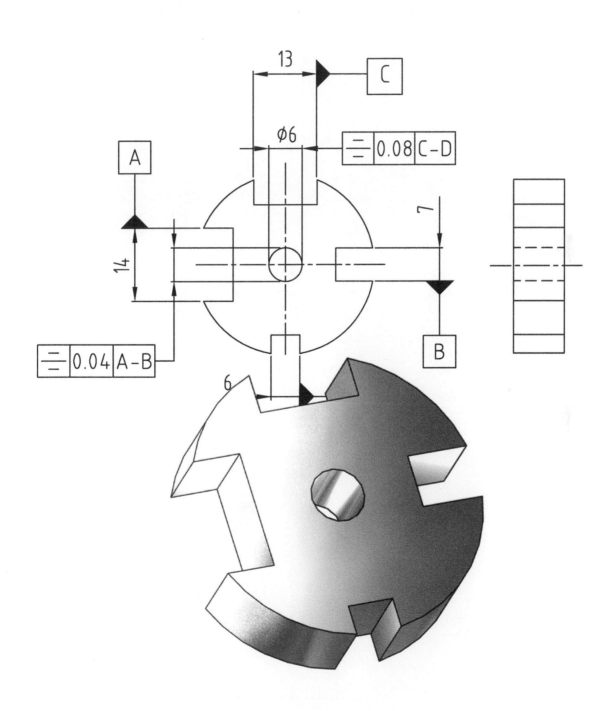

ANGULARITY ⟋

Description:

Defines the required condition of a | median plane or surface or axis | at a specific angle (except 90°)

from a datum | plane or axis |

	Form	Orientation	Run-out	Location	Median plane or Surface or Axis	Yes	No	Axis or Plane	Related	Single	True (exact position)
Tolerance type		✓									
What is controlled					✓						
Datum needed						✓					
What is the datum								✓			
For features that are									✓		
Notes: angle dimensions to be											✓

CIRCULAR RUN-OUT ⟋

Description:

Circular run-out defines how much an actual │ circular element │ can deviate from the required form (and orientation) during one complete revolution about the datum │ axis │

	Form	Orientation	Run-out	Location	Circular element	Yes	No	Axis	Related	Single	
Tolerance type			✓								
What is controlled					✓						
Datum needed						✓					
What is the datum								✓			
For features that are									✓		
Notes:											

COAXIALITY ◎

Description:

Controls the deviation of an | axis | from its true (exact) | axis |

	Form	Orientation	Run-out	Location	Axis	Yes	No (when implied datum is used, now obsolete)	Point or Axis	Related	Single	
Tolerance type				✓							
What is controlled					✓						
Datum needed						✓	*				
What is the datum								✓			
For features that are									✓		
Notes:											

CONCENTRICITY ◎

Description:

Controls the deviation of a | point | from its true (exact) | point |

	Form	Orientation	Run-out	Location	Point	Yes	No	Point	Related	Single
Tolerance type				✓						
What is controlled					✓					
Datum needed						✓				
What is the datum								✓		
For features that are									✓	
Notes:										

CYLINDRICITY

Description:

Cylindricity is the combination of straightness, roundness and parallelism when applied to a │ surface │

	Form	Orientation	Run-out	Location	Surface	Yes	No	No datum needed	Related	Single	Cylindrical
Tolerance type	✓										
What is controlled					✓						
Datum needed							✓				
What is the datum								✓			
For features that are										✓	
Notes: applies only to forms that are											✓

FLATNESS $\diagup\!\!\!\!\diagup$

Description:

Flatness controls the divergence of a │ surface │ from a true │ plane │

	Form	Orientation	Run-out	Location	Surface	Yes	No	No datum needed	Related	Single	
Tolerance type	✓										
What is controlled					✓						
Datum needed							✓				
What is the datum								✓			
For features that are										✓	
Notes:											

PARALLELISM //

Description:

Defines the required condition of a | surface or axis | to be a uniform distance from a | plane or axis |

	Form	Orientation	Run-out	Location	Surface or Axis	Yes	No	Axis or Plane	Related	Single	
Tolerance type		✓									
What is controlled					✓						
Datum needed						✓					
What is the datum								✓			
For features that are									✓		
Notes:											

PERPENDICULARITY ⊥

Description:

Defines the required condition of a | median plane or surface or axis | to be exactly 90° from a | plane or axis |

	Form	Orientation	Run-out	Location	Median plane or Surface or Axis	Yes	No	Plane or Axis	Related	Single
Tolerance type		✓								
What is controlled					✓					
Datum needed						✓				
What is the datum								✓		
For features that are									✓	
Notes:										

POSITION

Description:

Position limits the deviation of the position of a $\left|\begin{array}{c}\text{surface}\\\text{or}\\\text{axis}\\\text{or}\\\text{point}\\\text{or}\\\text{line}\end{array}\right|$ from a true position

	Form	Orientation	Run-out	Location	Surface or Axis or Point or Line	Yes	No	Axis or Point or Line or Plane	Related	Single	True (exact position)
Tolerance type				✓							
What is controlled					✓						
Datum needed						✓	✓				
What is the datum								✓			
For features that are									✓	✓	
Notes: position dimensions to be											✓

PROFILE OF A LINE ⌒

Description:

Controls the perfect form of a │ contour │ of a │ profile │

	Form	Orientation	Run-out	Location	Profile of a line	Yes (when controlling orientation or location)	No (when controlling form)	Axis or Point or Line or Plane	Related (orientation or location)	Single (form)	True (exact position)
Tolerance type	✓	✓		✓							
What is controlled					✓						
Datum needed						✓	✓				
What is the datum								✓			
For features that are									✓	✓	
Notes: contour dimensions to be											✓

PROFILE OF A SURFACE ⌓

Description:

Controls the perfect form of a │ surface │

	Form	Orientation	Run-out	Location	Surface	Yes (when controlling orientation or location)	No (when controlling form)	Axis or Point or Line or Plane	Related (orientation or location)	Single (form)	True (exact position)
Tolerance type	✓	✓		✓							
What is controlled					✓						
Datum needed						✓	✓				
What is the datum								✓			
For features that are									✓	✓	
Notes: profile dimensions to be											✓

ROUNDNESS ⌀

Description:

Controls the errors of form of a | line or circumference or periphery at section | in the | cross-section or circumferential | plane

	Form	Orientation	Run-out	Location	Line or Circumference or Periphery	Yes	No	No datum needed	Related	Single	
Tolerance type	✓										
What is controlled					✓						
Datum needed							✓				
What is the datum								✓			
For features that are										✓	
Notes:											

STRAIGHTNESS $\boxed{-}$

Description:

Defines the required condition of a $\left| \begin{array}{c} \text{line} \\ \text{or} \\ \text{axis} \\ \text{or} \\ \text{edge} \\ \text{or} \\ \text{surface element} \end{array} \right|$ to be a straight line

	Form	Orientation	Run-out	Location	Line or Axis or Edge or Surface element	Yes	No	No datum needed	Related	Single
Tolerance type	✓									
What is controlled					✓					
Datum needed							✓			
What is the datum								✓		
For features that are										✓
Notes:										

SYMMETRY ⊟

Description:

Defines the required condition of a | line or axis or median plane | to be positioned symetrically in

relation to a datum | line or median plane |

	Form	Orientation	Run-out	Location	Line or Axis or Median plane	Yes	No (when implied datum is used, now obsolete)	Line or Median plane	Related	Single	
Tolerance type				✓							
What is controlled					✓						
Datum needed						✓	*				
What is the datum								✓			
For features that are									✓		
Notes:											

SYMMETRY

TOTAL RUN-OUT

Description:

Total run-out defines how much an actual │ surface │ can deviate from the required form

(and orientation) during one complete revolution about the datum │ axis │

	Form	Orientation	Run-out	Location	Surface	Yes	No	Axis	Related	Single
Tolerance type			✓							
What is controlled					✓					
Datum needed						✓				
What is the datum								✓		
For features that are									✓	
Notes:										

Datum quick reference

Characteristic		Symbol	Datum needed Yes ✓ No ✗
ORIENTATION	Angularity	∠	✓
RUN-OUT	Circular run-out	↗	✓
LOCATION	Coaxiality	◎	✓
LOCATION	Concentricity	◎	✓
FORM	Cylindricity	⌭	✗
FORM	Flatness	▱	✗
ORIENTATION	Parallelism	//	✓
ORIENTATION	Perpendicularity (Squareness)	⊥	✓
LOCATION	Position	⊕	✓ or ✗
FORM	Profile of a line	⌒	✗
ORIENTATION	Profile of a line	⌒	✓
LOCATION	Profile of a line	⌒	✓
FORM	Profile of a surface	⌓	✗
ORIENTATION	Profile of a surface	⌓	✓
LOCATION	Profile of a surface	⌓	✓
FORM	Roundness (Circularity)	○	✗
FORM	Straightness	—	✗
LOCATION	Symmetry	≐	✓
RUN-OUT	Total run-out	⌰	✓

PART 5

Indexes

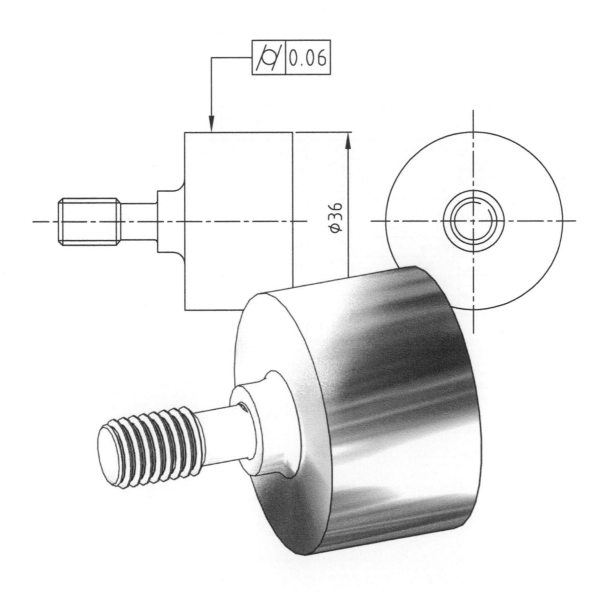

VISUAL INDEX 21

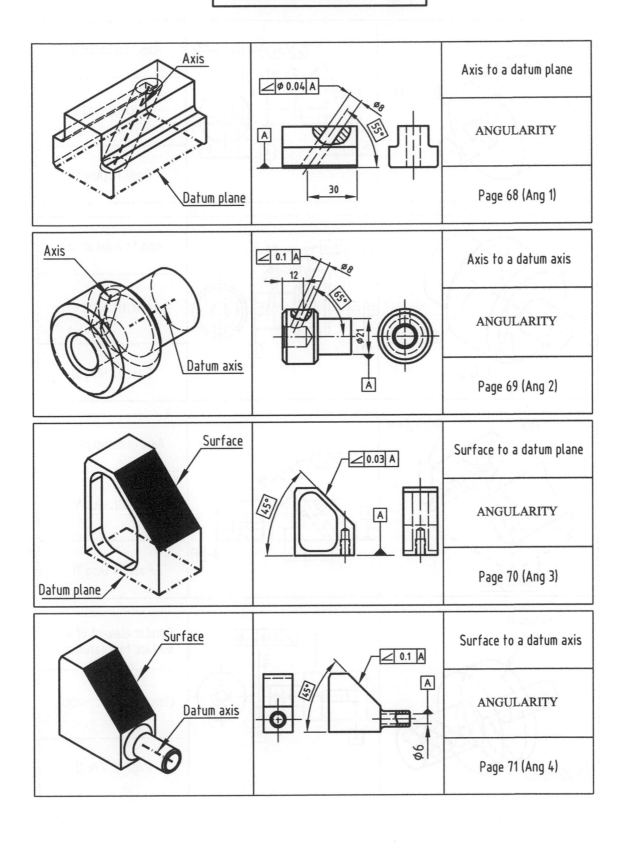

		Axis to a datum plane
		ANGULARITY
		Page 68 (Ang 1)
		Axis to a datum axis
		ANGULARITY
		Page 69 (Ang 2)
		Surface to a datum plane
		ANGULARITY
		Page 70 (Ang 3)
		Surface to a datum axis
		ANGULARITY
		Page 71 (Ang 4)

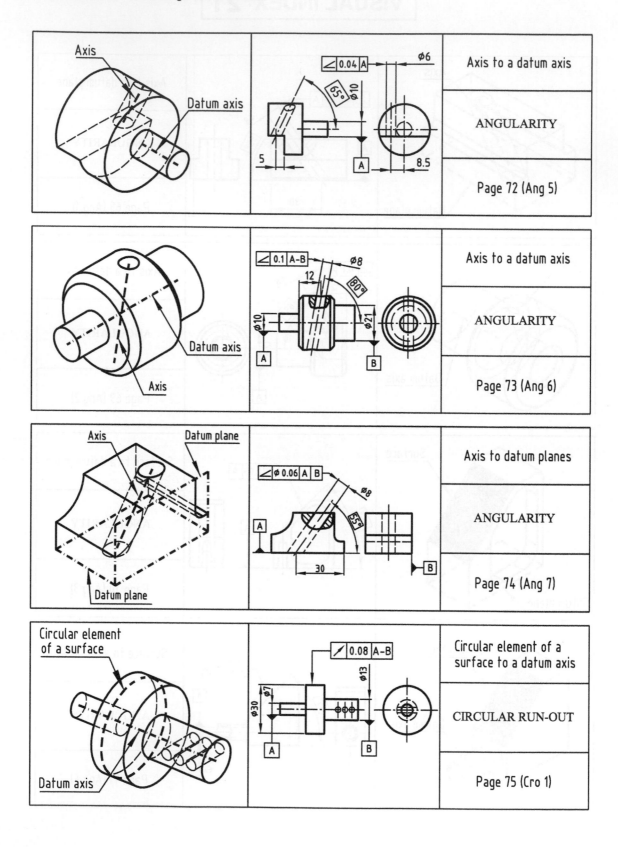

Axis / Datum axis	⌔ 0.04 A ⌀6 65° ⌀10 5 A 8.5	Axis to a datum axis ANGULARITY Page 72 (Ang 5)
Datum axis / Axis	∠ 0.1 A-B ⌀8 12 80° ⌀10 ⌀21 A B	Axis to a datum axis ANGULARITY Page 73 (Ang 6)
Axis / Datum plane / Datum plane	∠ ⌀0.06 A B ⌀8 55° A 30 B	Axis to datum planes ANGULARITY Page 74 (Ang 7)
Circular element of a surface / Datum axis	↗ 0.08 A-B ⌀13 ⌀7 ⌀30 A B	Circular element of a surface to a datum axis CIRCULAR RUN-OUT Page 75 (Cro 1)

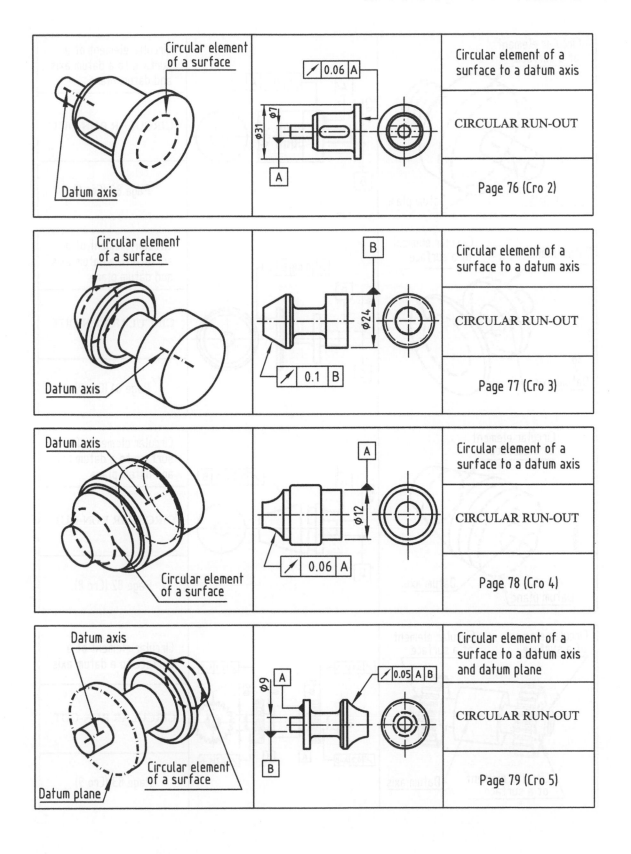

		Circular element of a surface to a datum axis
	0.06 A	CIRCULAR RUN-OUT
Circular element of a surface / Datum axis	Ø31 / Ø7 / A	Page 76 (Cro 2)
		Circular element of a surface to a datum axis
Circular element of a surface / Datum axis	B / Ø24 / 0.1 B	CIRCULAR RUN-OUT
		Page 77 (Cro 3)
		Circular element of a surface to a datum axis
Datum axis / Circular element of a surface	A / Ø12 / 0.06 A	CIRCULAR RUN-OUT
		Page 78 (Cro 4)
		Circular element of a surface to a datum axis and datum plane
Datum axis / Circular element of a surface / Datum plane	Ø9 / A / 0.05 A B / B	CIRCULAR RUN-OUT
		Page 79 (Cro 5)

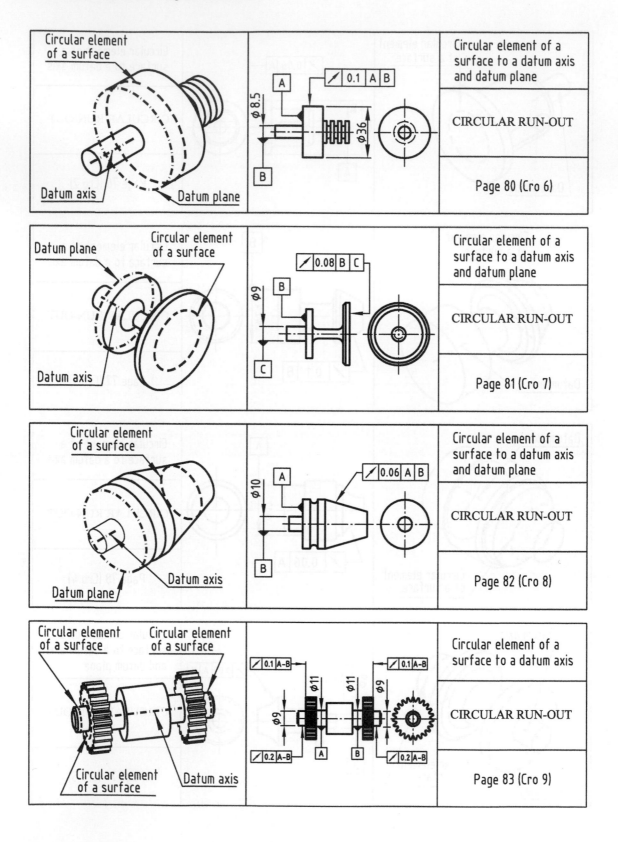

		Circular element of a surface to a datum axis and datum plane
		CIRCULAR RUN-OUT
		Page 80 (Cro 6)
		Circular element of a surface to a datum axis and datum plane
		CIRCULAR RUN-OUT
		Page 81 (Cro 7)
		Circular element of a surface to a datum axis and datum plane
		CIRCULAR RUN-OUT
		Page 82 (Cro 8)
		Circular element of a surface to a datum axis
		CIRCULAR RUN-OUT
		Page 83 (Cro 9)

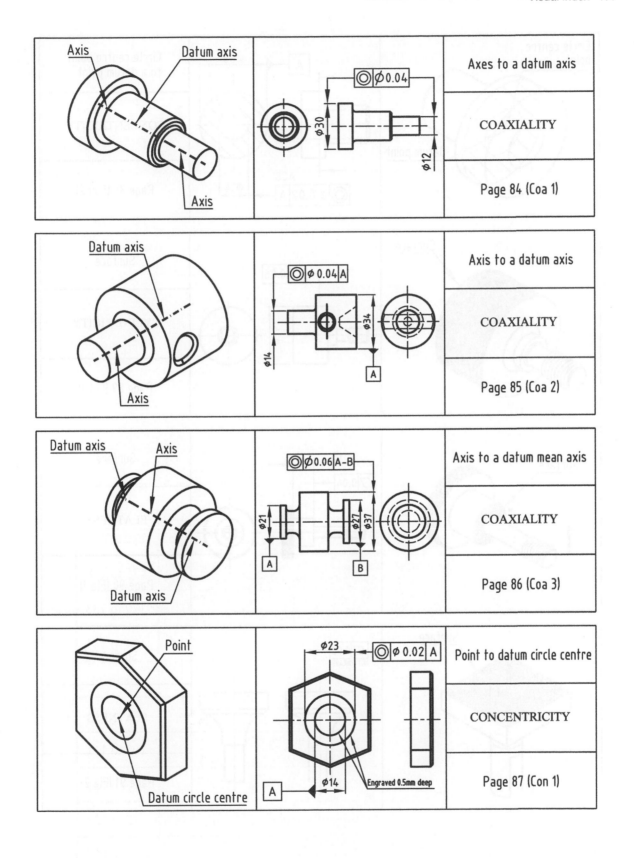

		Axes to a datum axis
		COAXIALITY
		Page 84 (Coa 1)
		Axis to a datum axis
		COAXIALITY
		Page 85 (Coa 2)
		Axis to a datum mean axis
		COAXIALITY
		Page 86 (Coa 3)
		Point to datum circle centre
		CONCENTRICITY
		Page 87 (Con 1)

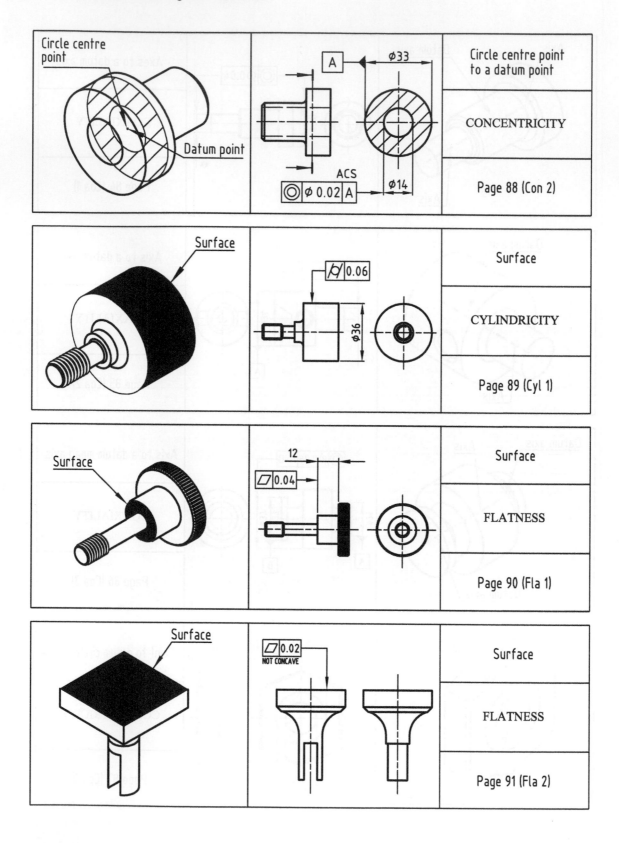

		Circle centre point to a datum point
Circle centre point / Datum point	A / ⌀33 / ⌀14 / ACS / ◎ ⌀ 0.02 A	CONCENTRICITY
		Page 88 (Con 2)
Surface	⌀/ 0.06 / ⌀36	Surface
		CYLINDRICITY
		Page 89 (Cyl 1)
Surface	12 / ▱ 0.04	Surface
		FLATNESS
		Page 90 (Fla 1)
Surface	▱ 0.02 / NOT CONCAVE	Surface
		FLATNESS
		Page 91 (Fla 2)

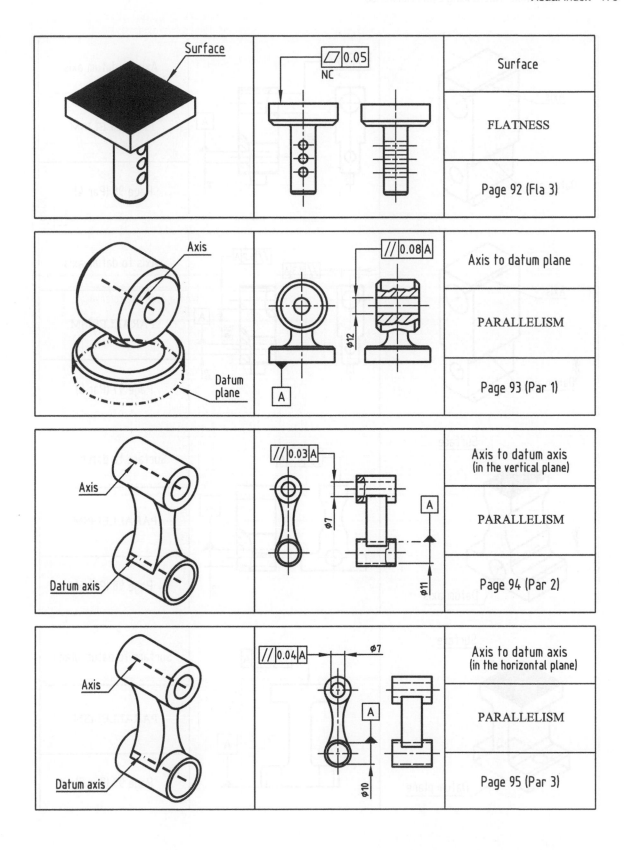

		Surface
		FLATNESS
		Page 92 (Fla 3)
		Axis to datum plane
		PARALLELISM
		Page 93 (Par 1)
		Axis to datum axis (in the vertical plane)
		PARALLELISM
		Page 94 (Par 2)
		Axis to datum axis (in the horizontal plane)
		PARALLELISM
		Page 95 (Par 3)

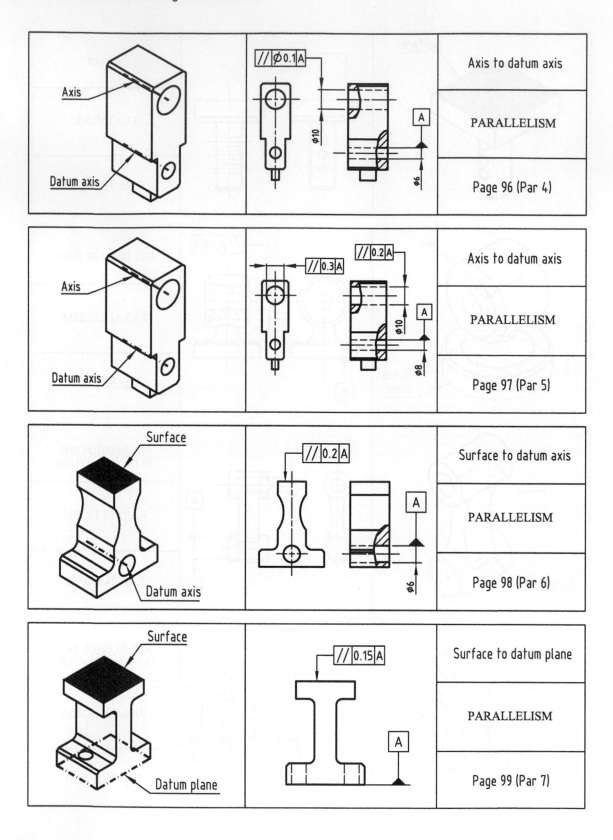

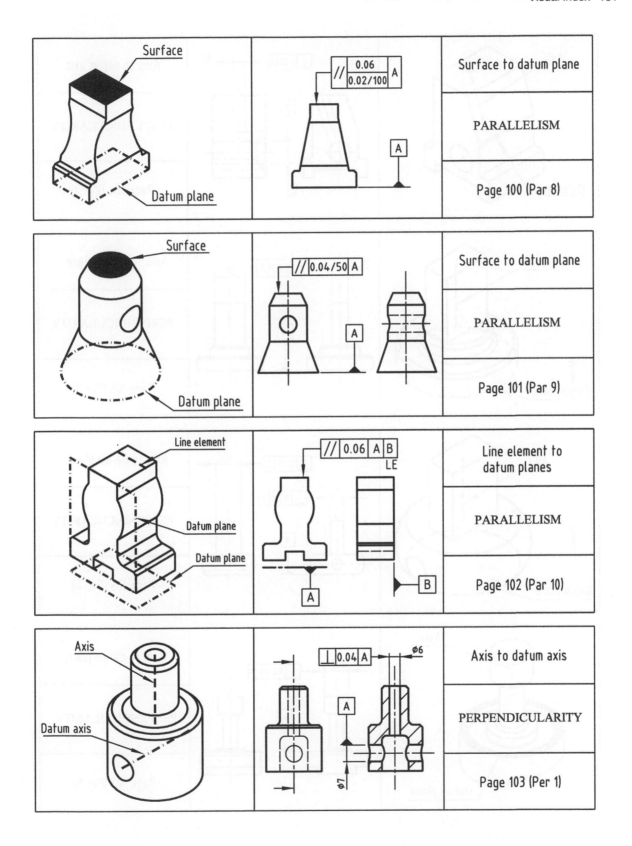

		Surface to datum plane
		PARALLELISM
		Page 100 (Par 8)
		Surface to datum plane
		PARALLELISM
		Page 101 (Par 9)
		Line element to datum planes
		PARALLELISM
		Page 102 (Par 10)
		Axis to datum axis
		PERPENDICULARITY
		Page 103 (Per 1)

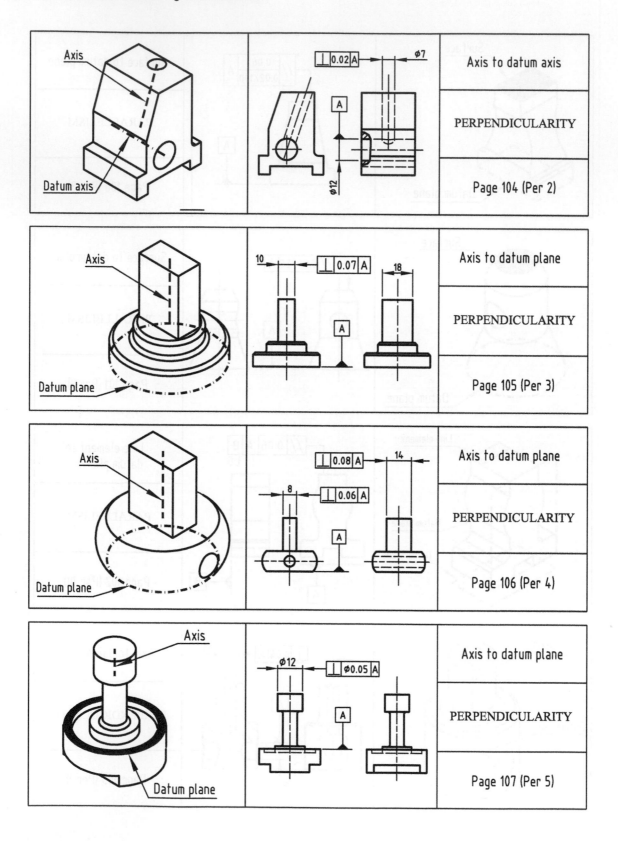

Axis / Datum axis	⊥ 0.02 A / A / ⌀7 / ⌀12	Axis to datum axis PERPENDICULARITY Page 104 (Per 2)
Axis / Datum plane	10 / ⊥ 0.07 A / 18 / A	Axis to datum plane PERPENDICULARITY Page 105 (Per 3)
Axis / Datum plane	⊥ 0.08 A / 14 / 8 / ⊥ 0.06 A / A	Axis to datum plane PERPENDICULARITY Page 106 (Per 4)
Axis / Datum plane	⌀12 / ⊥ ⌀0.05 A / A	Axis to datum plane PERPENDICULARITY Page 107 (Per 5)

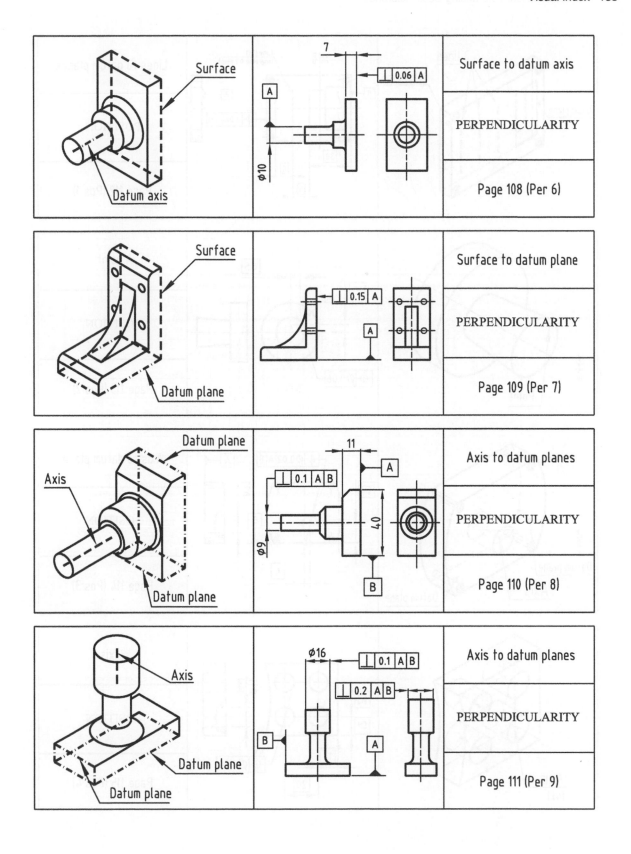

		Surface to datum axis
		PERPENDICULARITY
		Page 108 (Per 6)
		Surface to datum plane
		PERPENDICULARITY
		Page 109 (Per 7)
		Axis to datum planes
		PERPENDICULARITY
		Page 110 (Per 8)
		Axis to datum planes
		PERPENDICULARITY
		Page 111 (Per 9)

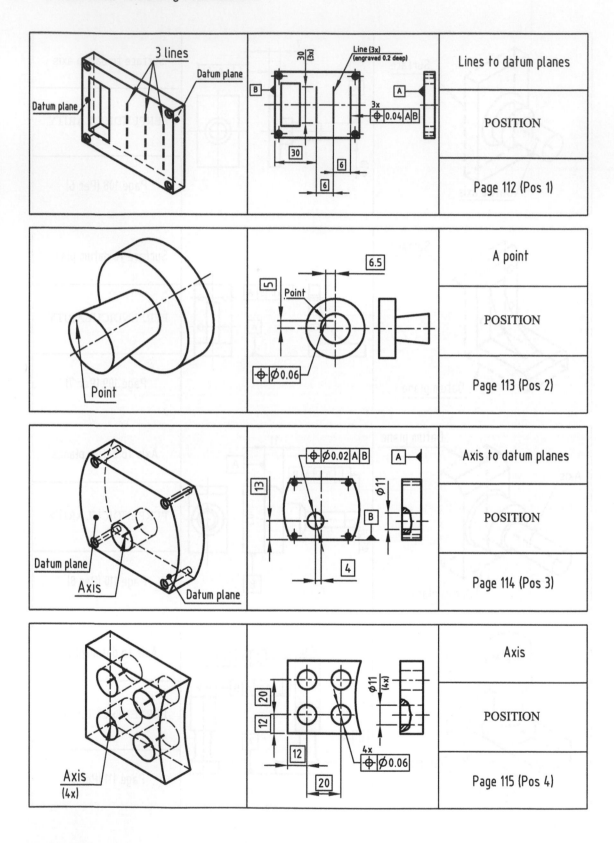

		Lines to datum planes
3 lines · Datum plane · Datum plane	Line (3x) (engraved 0.2 deep) · 30 (3x) · B · A · 3x ⊕ 0.04 A B · 30 · 6 · 6	POSITION
		Page 112 (Pos 1)

		A point
Point	6.5 · 5 · Point · ⊕ ∅0.06	POSITION
		Page 113 (Pos 2)

		Axis to datum planes
Datum plane · Axis · Datum plane	⊕ ∅0.02 A B · A · 13 · ∅11 · B · 4	POSITION
		Page 114 (Pos 3)

		Axis
Axis (4x)	20 · 12 · 12 · ∅11 (4x) · 4x ⊕ ∅0.06 · 20	POSITION
		Page 115 (Pos 4)

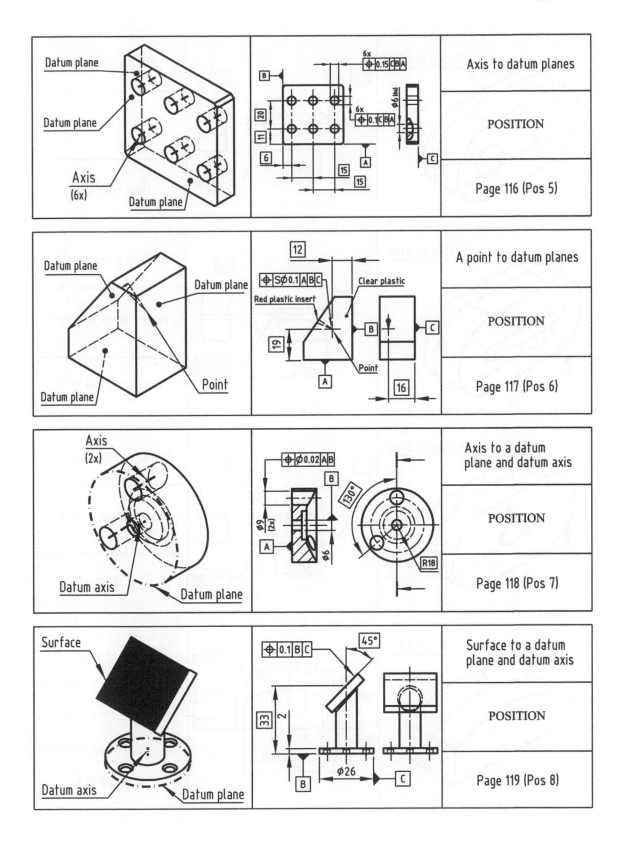

		Axis to datum planes
		POSITION
		Page 116 (Pos 5)
		A point to datum planes
		POSITION
		Page 117 (Pos 6)
		Axis to a datum plane and datum axis
		POSITION
		Page 118 (Pos 7)
		Surface to a datum plane and datum axis
		POSITION
		Page 119 (Pos 8)

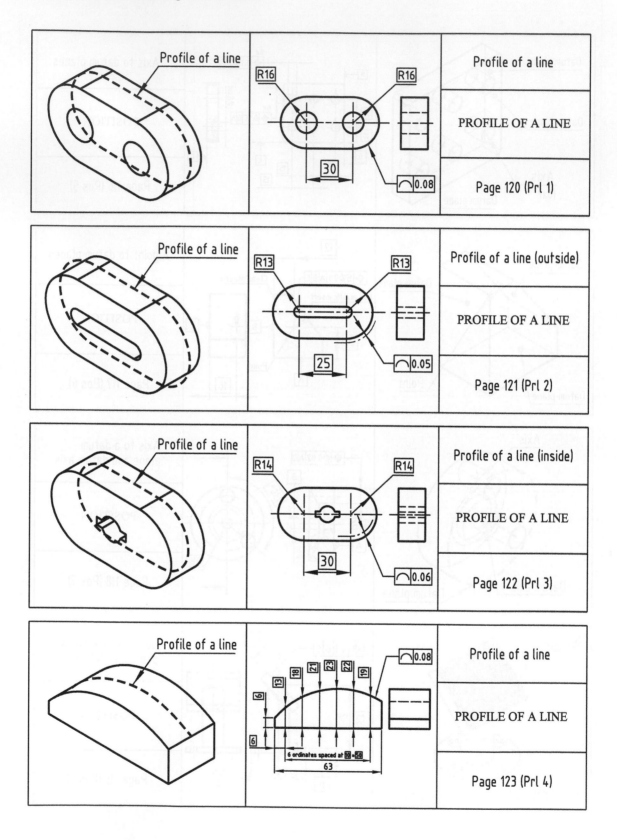

		Profile of a line
Profile of a line	R16 R16 30 ⌒0.08	PROFILE OF A LINE
		Page 120 (Prl 1)
Profile of a line	R13 R13 25 ⌒0.05	Profile of a line (outside)
		PROFILE OF A LINE
		Page 121 (Prl 2)
Profile of a line	R14 R14 30 ⌒0.06	Profile of a line (inside)
		PROFILE OF A LINE
		Page 122 (Prl 3)
Profile of a line	⌒0.08 6 13 18 21 23 22 19 6 6 ordinates spaced at 10=50 63	Profile of a line
		PROFILE OF A LINE
		Page 123 (Prl 4)

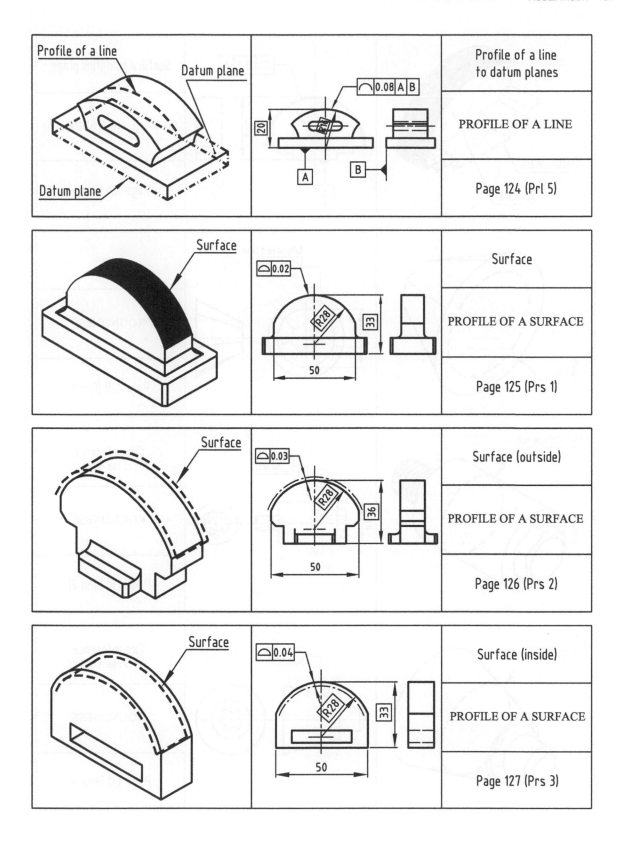

		Profile of a line to datum planes
Profile of a line / Datum plane / Datum plane	⌒ 0.08 A B / 20 / A / B	PROFILE OF A LINE
		Page 124 (Prl 5)
		Surface
Surface	⌓ 0.02 / R28 / 33 / 50	PROFILE OF A SURFACE
		Page 125 (Prs 1)
		Surface (outside)
Surface	⌓ 0.03 / R28 / 36 / 50	PROFILE OF A SURFACE
		Page 126 (Prs 2)
		Surface (inside)
Surface	⌓ 0.04 / R28 / 33 / 50	PROFILE OF A SURFACE
		Page 127 (Prs 3)

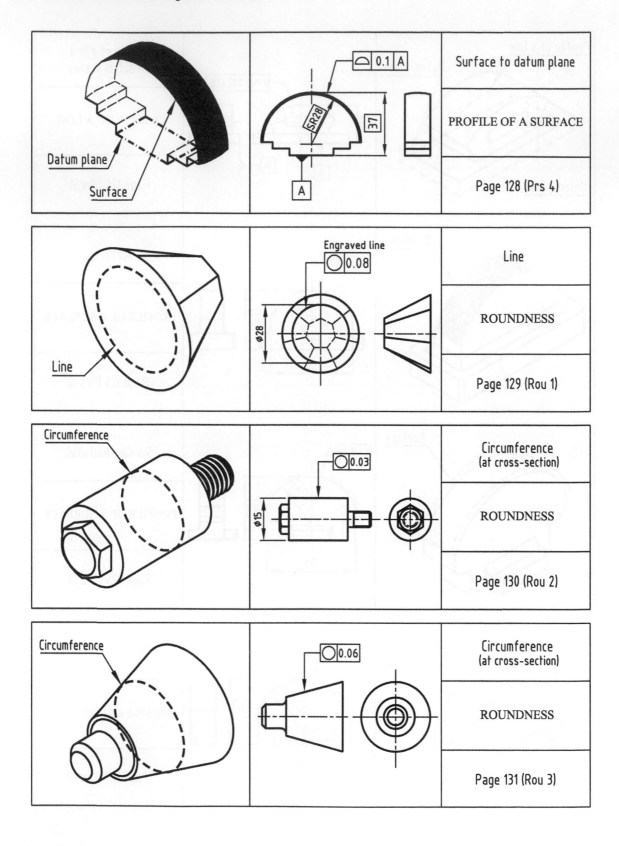

		Surface to datum plane
Datum plane / Surface	△ 0.1 A / SR28 / 37 / A	PROFILE OF A SURFACE
		Page 128 (Prs 4)

	Engraved line / ○ 0.08 / Ø28	Line
Line		ROUNDNESS
		Page 129 (Rou 1)

Circumference	○ 0.03 / Ø15	Circumference (at cross-section)
		ROUNDNESS
		Page 130 (Rou 2)

Circumference	○ 0.06	Circumference (at cross-section)
		ROUNDNESS
		Page 131 (Rou 3)

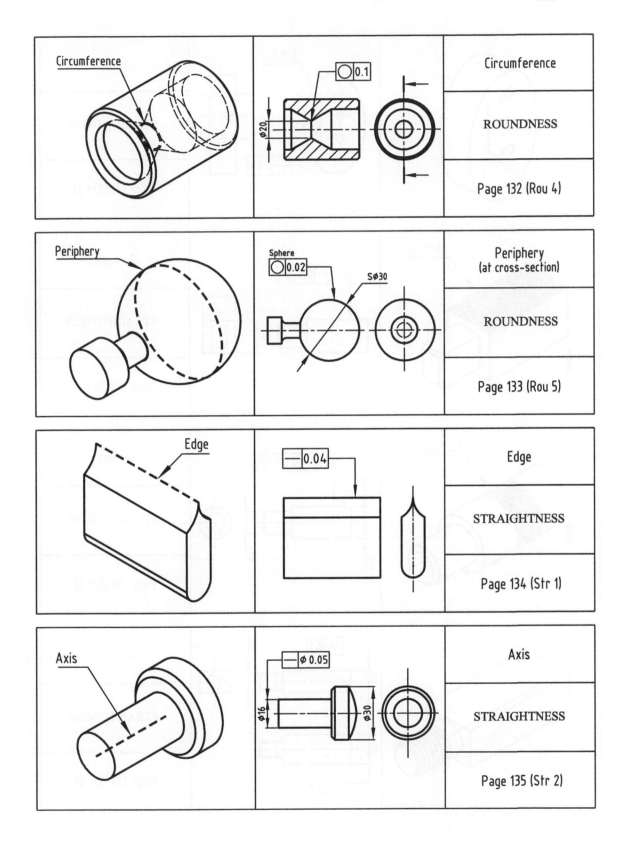

		Circumference
		ROUNDNESS
		Page 132 (Rou 4)
		Periphery (at cross-section)
		ROUNDNESS
		Page 133 (Rou 5)
		Edge
		STRAIGHTNESS
		Page 134 (Str 1)
		Axis
		STRAIGHTNESS
		Page 135 (Str 2)

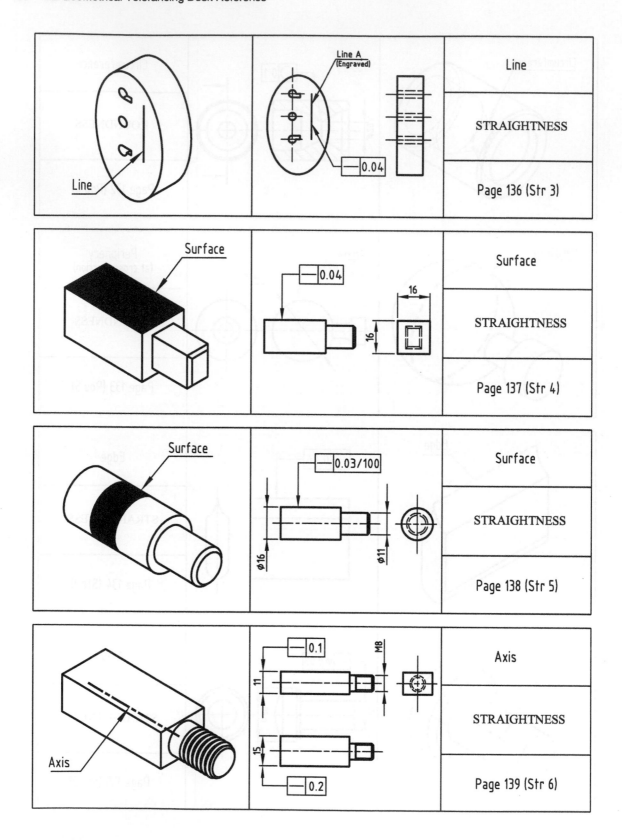

		Line
		STRAIGHTNESS
		Page 136 (Str 3)
		Surface
		STRAIGHTNESS
		Page 137 (Str 4)
		Surface
		STRAIGHTNESS
		Page 138 (Str 5)
		Axis
		STRAIGHTNESS
		Page 139 (Str 6)

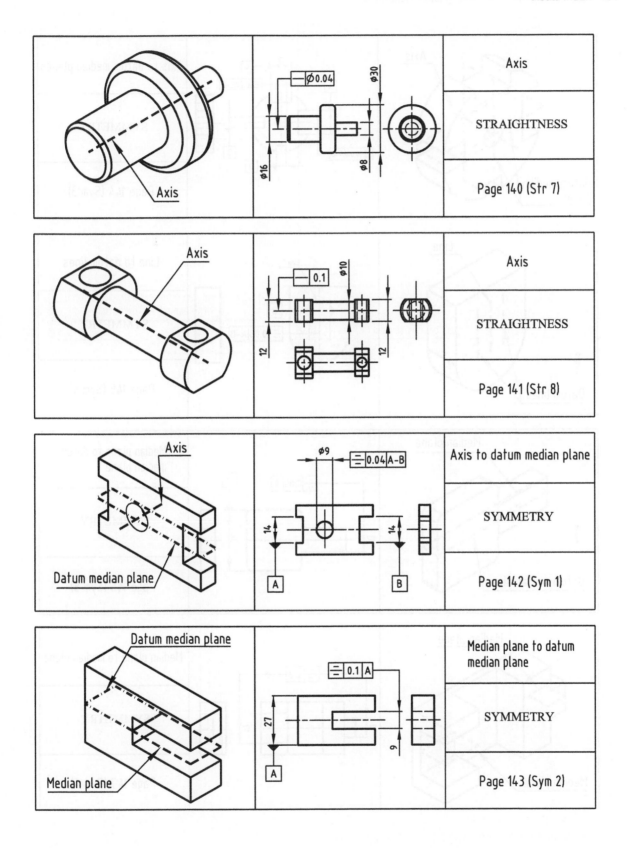

		Axis
	Ø0.04 Ø30 Ø16 Ø8	STRAIGHTNESS
Axis		Page 140 (Str 7)
Axis		Axis
	0.1 Ø10 12 12	STRAIGHTNESS
		Page 141 (Str 8)
Axis Datum median plane	Ø9 0.04 A-B 14 14 A B	Axis to datum median plane SYMMETRY Page 142 (Sym 1)
Datum median plane Median plane	0.1 A 27 9 A	Median plane to datum median plane SYMMETRY Page 143 (Sym 2)

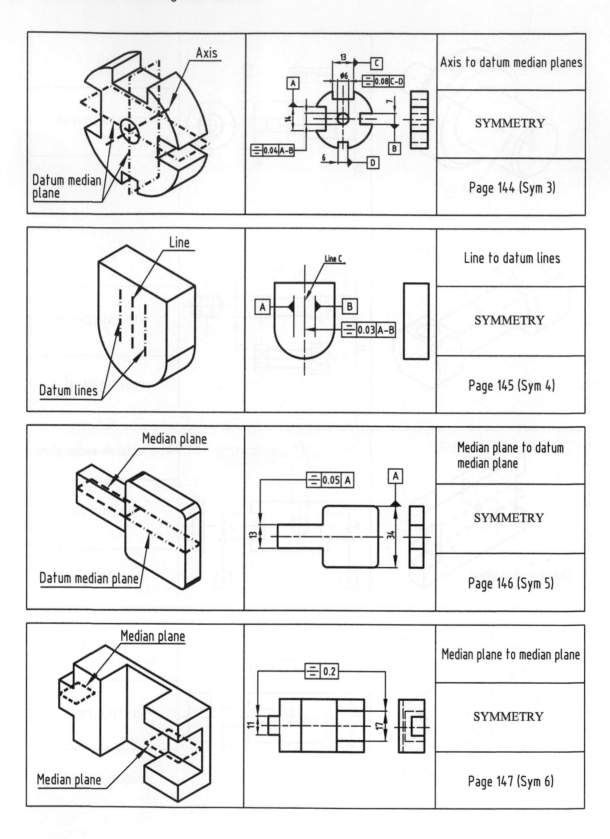

		Axis to datum median planes
		SYMMETRY
		Page 144 (Sym 3)
		Line to datum lines
		SYMMETRY
		Page 145 (Sym 4)
		Median plane to datum median plane
		SYMMETRY
		Page 146 (Sym 5)
		Median plane to median plane
		SYMMETRY
		Page 147 (Sym 6)

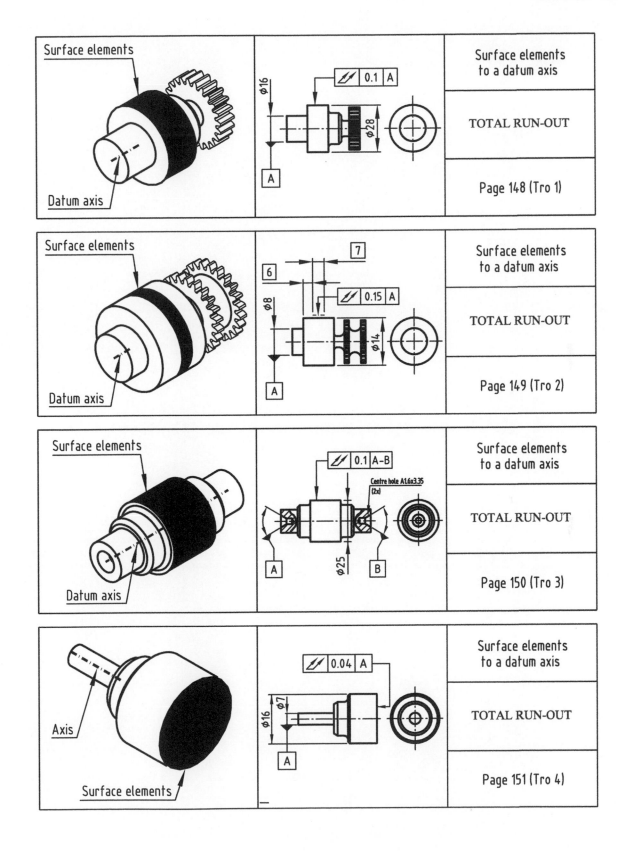

		Surface elements to a datum axis
Surface elements / Datum axis	Ø16 / ⫽ 0.1 A / Ø28 / A	TOTAL RUN-OUT
		Page 148 (Tro 1)

		Surface elements to a datum axis
Surface elements / Datum axis	6 / 7 / Ø8 / ⫽ 0.15 A / Ø14 / A	TOTAL RUN-OUT
		Page 149 (Tro 2)

		Surface elements to a datum axis
Surface elements / Datum axis	⫽ 0.1 A–B / Centre hole A1.6x3.35 (2x) / Ø25 / A / B	TOTAL RUN-OUT
		Page 150 (Tro 3)

		Surface elements to a datum axis
Axis / Surface elements	⫽ 0.04 A / Ø7 / Ø16 / A	TOTAL RUN-OUT
		Page 151 (Tro 4)

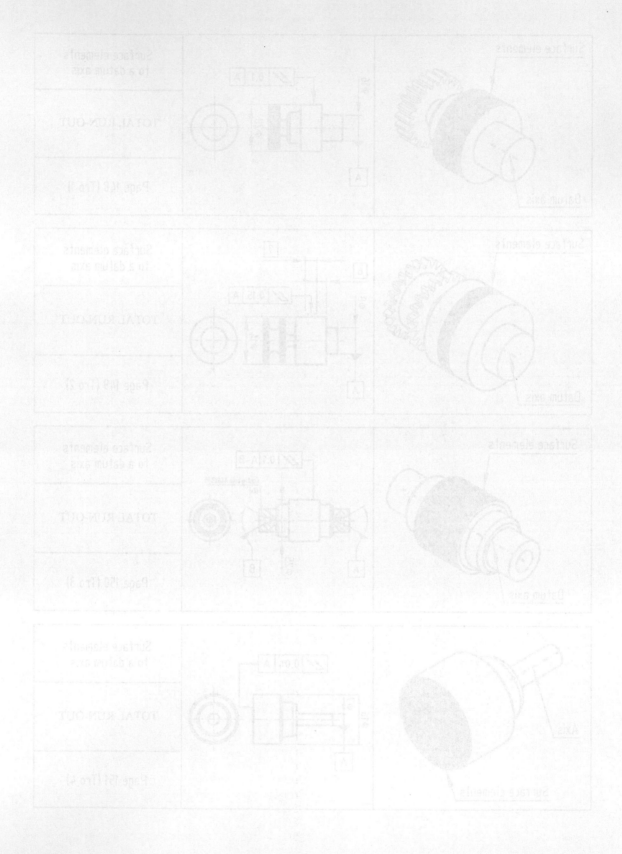

Printed and bound by CPI Group (UK) Ltd, Croydon, CR0 4YY

03/10/2024

01040336-0020